ANNE

UNPLUGGED
from the norm

How a life became a mission

Unless otherwise indicated, Bible quotations are taken from The King James Version of the Bible (KJV), available in the Public Domain.

UNPLUGGED
FROM THE NORM

By Anne Ranta

Copyright © 2013 by Anne Ranta

All rights reserved. This book is protected under the copyright laws of the United States of America. No part of this book may be reproduced in any form without written permission from the publisher, except in case of brief quotations in critical articles or reviews. This book may not be copied or reprinted for commercial gain or profit. The use of short quotations or occasional page copying for personal or group study is permitted.

If you would like to contact Anne Ranta or learn more about her charitable activities, please visit: www.HealingHeartsBalkans.org

This book may be purchased in paperback directly from the author by writing to: order.unplugged@gmail.com Also available online both in paperback and in formats for electronic readers.

*We idolize
wealth, fame, power;
we are a celebrity culture.
Just look at the movies we see,
the magazines we read. We don't
feature homeless people, ugly or
even ordinary-looking people.
Yet, Jesus
held up the poor, the persecuted,
those who mourn, as blessed,
as heirs of the kingdom.
We're called
to see the world with
different eyes, God's eyes,
and then to present that vision
in a convincing way.*

— Phillip Yancey

INTRODUCTION

Rick Warren, author of the best-selling book, *The Purpose Driven Life*, has said, "The purpose of influence is to speak up for those who have no influence." I have made my life's mission to stand with those very people and to do whatever I can to share God's love with them, to give to the needy and show compassion to the broken-hearted and those who have suffered loss. How much voice does a person have who supports himself and his family by scrounging daily through the city's dumpsters?

When Jesus called his disciples, He told them, "Follow Me, and I will make you fishers of men" (Matthew 4:19). And when they worried about how they would survive and "make a living," he taught them on the mount that they should "Lay not up for yourselves treasures on earth," and "Take no thought for the morrow," for the Father would feed them like the birds of the air, and that if they would "Seek first the Kingdom of God, then all these things would be added unto them" (Matthew 6:19-33). This is one of my favorite passages of Scripture and one that I believed could be lived out today. But how?

Were such promises written only for the early disciples? Did God's power to "supply our every need according to His riches in glory" (Philippians 4:19) dissipate over the years so that now we must depend on jobs, employers, investments, and the occasional win at the lottery in order to survive? I didn't think so, and I was determined to put God's Word to the test.

The following chapters document only a tiny fraction of my journey over the past 40 years of living by faith, along with my husband Andrew, in the truth that if one believes and trusts in these promises, they are as real as they were in 30 A.D. I was determined to step out onto that branch and then saw it off. I wanted to see how God would catch me and take care of me. By reading the stories and testimonies in this book, you too will be able to experience that thrill.

In John 8:32, Jesus said, "Ye shall know the truth, and the truth shall make you free." That truth for me was that God's promises are real, and His supply is unfailing, as long as we appropriate those promises and obey what He has said to do. "Go into all the world and preach the gospel" (Matthew 16:15). It has nothing to do with religion, our righteousness, our success or accomplishments. It has to do with simple faith in His Word. Nothing more.

So, how did I manage to live for forty years without a fixed income or a "normal" job? How could I pay for airfares or transatlantic cruises? How could I pay the hospital bill following a life-threatening ectopic pregnancy and ensuing operation? How could I educate and feed my children? Simple! By "seeking first the Kingdom of God."

We have never been rich by worldly standards, but, to be honest, we have felt as if we are millionaires! We have enjoyed, for the most part, good health, happiness, and inner peace—things that money cannot buy. During our lives of service, we have traveled to many exciting and exotic places, such as Guatemala and Belize in Central America; Peru, Argentina, and Uruguay in South America; India, Pakistan, Bangladesh, and the Himalaya Mountains in Asia; Jordan, Israel, and Palestine in the Mideast; and Libya, Tunisia, and Malta in North Africa, just to mention a few and all because of our desire to help the needy! The cost of all this traveling—where did that money come from? It was all supplied by supernatural means, or you could say "natural means" if you are heavenly-minded and a citizen of the Kingdom of God, for we have a very rich Father who owns "the cattle upon a thousand hills" (Psalm 50:10).

One person whose testimonies of faith greatly inspired us to live such a life was George Müller! Müller worked with orphans in Bristol, England, in the mid-1800s and built five homes that housed over 1,700 children. The cost of the original structures, along with the money needed to maintain them and feed and clothe the children, all came via donations, but he never made requests for financial support, nor did he go into debt. God knew his needs, and Müller looked only to his Heavenly Father to supply them.

Like George Müller, once we made the decision to dedicate our lives to God's service, it was like the Bank of Heaven issued us a Gold Card that we could swipe any time there was a need, and the tickets would just fall from the heavenly vaults!

I want to encourage everyone reading, especially the younger generation, that our God is truly an awesome God and that what He has done for us, He will do for you if you trust Him and commit your ways to Him (Psalm 37:5).

CHAPTER ONE

God's Insurance Policy

W orking with young people brings back memories of when I myself was lost, lonely, and seeking identity and purpose in life.

I am continually amazed at how the Lord has taken care of me my entire life. It is almost as if I have had some sort of divine insurance policy!

I was born and raised in the safety and security of Scandinavia. Sadly, I did not appreciate how blessed I was. In recent years, living in developing countries, I have seen so many young

Anne as a child

people longing to go to a more advanced, western country. In my youth, I was also discontent with my reality and dreamed of adventure in far-off lands. I distinctly remember attending university on the occasion of my 21st birthday—a major milestone in my life. It was then that I started to seriously consider my future. Would I become a social anthropologist or an astronomer? Or would I possibly be an architect or a computer programmer (at the time when a computer could fill an entire room). In my heart, I did not want to dedicate my life to materialistic pursuits, but neither did I see any long-term future happiness in the hippie culture that was popular among the youth of those days.

My father, of course, wanted me to stick with my university studies and work hard in order to guarantee myself a good, comfortable future. My professors were certainly not my idea of role models. I wanted to do something *special* with my life, but I had no idea what. One thing I knew: I did not want to simply be a tool of what I called "the system." I wanted to be me, and dedicate my life to something that I could look back on years later and say, "I am proud of who I became and what I did with my life!" I wanted to change the world. I wanted to make a difference.

Around that time, an aunt that I had been very close to while growing up was nearing her retirement age. She had worked as a secretary for Nokia for forty years (back in the days when they were only making rubber boots) and was looking forward to her hard-earned pension and a comfortable retirement. In her final year, she was laid off for no apparent reason. The company did not give her a penny for her retirement. She was devastated, and I got my first good look at how the dog-eat-dog world operates and treats people.

I also had an uncle who took it upon himself to make sure I was being properly educated in politics, as he himself was the head of an influential political party in northern Finland. When I would come home from school, his first question to me would be, "So, what did they teach you today?" When I told him, he would proceed to explain to me how it was not that way at all, and "This is how it really happened!" The history books actually changed after I finished my schooling! A result of that, I learned from a very early age to question education as well as the media! I would not take things simply at face value. I wanted to investigate and discover the truth through my own experience.

I held many different summer jobs, working in a factory, a

hospital, as a telephone operator, and at building sites where I would mix with the manual laborers. I really liked being in their company as they had, in my opinion, a better, more well-rounded general education than most of my university friends. At university, I would meet only very narrow-minded people who could converse only about their area of expertise, so it was impossible to have a normal conversation with them.

I decided to study social anthropology, as I was fascinated by the Bedouin culture and was able to take an exploratory trip to Palestine. Now, finally, I could fulfill my dream of traveling to distant lands! I had my one and only hippie experience in the south of Israel by the Red Sea, where I hitchhiked through the desert, lived in beach shacks, was chased by the police. I even spent a week in protective custody of the Israeli Defense Forces. I had been hitchhiking alone (a blond foreign girl—*not* to be recommended!) and was picked up by the Israeli military who did not feel it was safe, especially since it was in Arab territory. I realized they felt it was for my "protection," but the problem was that they were not willing to let me go and insisted on sending me back to Jerusalem against my wishes. There was a filming crew from the BBC there that had come to make a special report on happenings in the area, and when I had a chance to explain my predicament to them, they promised the authorities that they would take me with them out of the country. Once we were out of sight, they let me go to continue my adventure. I was foolish to be doing all that alone, and I believe I had a legion of angels keeping me safe from all harm. As a result of my field trip, I got a real first-hand insight into the situation in Palestine and fell in love with the Arabs' sweet nature!

In spite of the amazing things I was able to experience during that time in Palestine, I fell into a deep discouragement as I anticipated the world and my life when the end of my stay was approaching. During my time in Jerusalem, I had met a young English man, and in spite of having known me for only a few months, he ended up proposing to me. Since I did not have any other clear direction for my life at that point, I figured why not?

We settled down in England. He had long, frizzy blond hair, something that would definitely *not* be to my father's liking! His brother-in-law was one of the performers in the hit musical "Hair" in London, so even though we were hippies at heart, we were able to hobnob with the "rich and famous" and experience their lifestyle as well! But after living the "normal" life for a while—working during the day and him watching football in the evenings—I, like Solomon of old, began to lament that all in life was vanity.

I had a very good job in a pathology laboratory where I, again, learned many life lessons, this time regarding the truth about doctors, medicines, and the medical/pharmaceutical industry. Laboratory technicians did not have a good attitude toward doctors! I was told that doctors most often guess when it comes to diagnosing patients, and the medicine they prescribe is often either by guesswork or by using the medicines that their friend, the drug rep, recommends. We in the laboratory knew what the problem was after our test results were analyzed, and only then should the correct medicine have been prescribed.

Since I had free health care while working in the hospital, I decided to have an operation done on my toe. When slipping away under the anesthesia, I felt the doctor examining my *knee*. I woke up by a miracle and told the doctor it was not my *knee* but my *toe*! Later, I learned that there was another patient with the same first name that had a knee problem. I am very thankful that I still have both my legs! This cured me for life from trusting any advice from doctors or the medicines they prescribe!

Even though I enjoyed my work in the laboratory, I was not about to make it my life calling. I decided to return to Sweden to study computers, thinking that this was the thing of the future and that perhaps making money was important after all. My husband wanted to return to Israel, so we were not together much after that and ended up divorcing a few years later.

I was accepted to study computers and mathematics, but I think in my heart I knew I was not making the right decision. After a few months, I was sitting one day in the university cafeteria when some foreign visitors came in and sang a few songs that were obviously Christian, but they did not come across as being churchy. It was so unusual that it appealed to me. When I was young, my grandmother who was raising me was very Christian. She would take me to church every Sunday, but the Finnish state church that she attended was very traditional and dead. As I grew up, I determined that I would rather be a communist than go to church. There was no way I would ever darken a church door! But these people seemed to have some sort of aura about them. I saw in them a freedom and happiness that I had been searching for. They were just simple, uncomplicated people who wanted to use their lives for a higher purpose and help the world. They were people who were willing to move out of their comfort zones to do something for others. I talked with them, and they explained to me how they were living like Jesus did in the Bible. They said He was a revolutionary who dared to differ from the status quo, who was a friend of harlots and sinners, and who was not concerned with His position in society. That was more my idea of what Jesus would be like. He was radically different from what most people would consider to be a "religious" person. He seemed to do everything contrary to natural expectations and was unwilling to conform to their religious laws.

Jesus had taught His disciples to "be like the birds" and not worry about food or clothing or shelter, which seemed so opposite to the natural ways. I wanted that extra energy he had to give, so I could be different. I hated "normal." I needed that miraculous manual to explain how to do it, God's "Cookbook of Life"! I got a Bible and started to study it. I spent half a year in isolation reading and applying the Word, after which I decided that I wanted to dedicate my life to this! I had met the Author of the book!

I never regretted that decision. I had found the most amazing life! I just needed to take a step out of my little world into the Kingdom of Jesus by accepting Him to rule my life. I certainly had made a mess of it myself.

Now, when I look back on my life, I have a great feeling of

satisfaction. How wonderful God has been to me! He certainly is the best insurance policy. He has taken such amazing care of me—my problems, sicknesses, my children and their education, and He has supplied all our needs in many difficult and poor mission fields. I certainly thank Him for giving me this wonderful life that has not been dependent on man's "social security" but on God's heavenly security!

For the past forty years, I have traveled around the world in search of other people who are lost and lonely like I once was and offering them a life with meaning and purpose. I will continue to do so until I meet my first Love, Jesus, face to face! There is no more fulfilling sight than that of a searching soul finding the answers to his or her life. There is nothing more satisfying than stepping out of our selfishness to help others.

CHAPTER TWO

The Adventure Begins

After meeting those Christian young people in the university cafeteria that day and listening to how they were following Jesus just like the twelve disciples of the early church, I knew what the Lord wanted me to do with my life. My search then turned to finding out where he wanted me, since I still felt called to foreign lands, and whether He would now supply for me a partner with whom I could work out His plan for my life.

I traveled down to Italy and then Sicily and ended up in Malta. It was there that I met Andrew. He had his own born-again experience the previous year after having become disillusioned with life in the U.S. following the Vietnam war debacle and had traveled to Europe to visit his brother and see what the Lord had planned for his life. We were like two jigsaw puzzle pieces that fit together perfectly. Once in place, the full picture started to become clear. Andrew had told me that during all those years in university and while he was working, he had never met anyone that he considered marrying. Once he was born again, he understood that the Lord had that special someone prepared for him. It was not long before we were married and the children began to come.

In praying about a mission field where we could begin our

Anne and Andrew newly engaged

service for Him together, we felt called to Jordan. I had already had some experience there, and English was widely spoken.

As mentioned previously, during my time studying the Bedouins I had developed a great love and respect for the Palestinian people, and we felt great compassion for them and their situation due to all the suffering they were enduring as a result of the politics of the region. We wanted the opportunity to show the Lord's love to them. And so it was that after spending a few months in France and Switzerland in preparation, and having received a small wedding gift from Andrew's mother, we purchased an old Peugeot 403 sedan for exactly $200 and set out together in our new life by faith.

The Apostle John says at the end of his gospel account, *"And there are also many other things which Jesus did, the which, if they should be written every one, I suppose that even the world itself could not contain the books that should be written"* (John 21:25 KJV). As much as I would like to write in this book a detailed account of each and every amazing thing that the Lord has done for us over the years, I'm afraid I will have to stick to what I feel are the major highlights in the hope that you will be strengthened and encouraged in your own faith walk as you read.

To us, Jordan was a dream country, close to where Jesus had spent his childhood! Although it was desert, to us it was beautiful as we looked at it through the eyes of those who called this land their home!

Children playing in Jordan

While there, we mostly worked with Palestinian refugees and volunteered at an orphanage and one of Mother Teresa's "Sisters of Charity" homes for the aged! Unfortunately, due to various circumstances, our time there was soon over, and we found the Lord leading us halfway around the world to Argentina.

Once we were sure that we were operating according to the Lord's plan for us, we had the confidence to put in our travel request with the Halls of Heaven. Greece would be the launching point for our trans-Atlantic crossing, but we first had to travel overland through Syria and Turkey with our French-registered car, our Jordanian-registered caravan that we had bought some months prior, two young children, and our dog. We expected some bumps in the road, as we knew that the mountains in Turkey would present a formidable obstacle to our heavily-laden rig, but our first hurdle came before we could even get started!

Border tensions had escalated between Jordan and Syria, as Syria was accusing Jordan of harboring Syrian opposition rebels, and then we heard that the border was totally closed. On top of that, U.S. relations with Syria were also strained, and the Syrian embassy in Amman refused to give Andrew even a 48-hour transit visa. Had we somehow missed the mark? Was this the enemy fighting our move? Or was it some sort of test? These are always questions that come up when serving the Lord and seeking His will.

The first answer came as I was in the ladies' room at the Syrian embassy crying out to the Lord for help in this seemingly hopeless situation. I walked out of the ladies' room with my eyes reddened from crying, and a high-level official saw me and asked what the problem was. After a brief explanation, he asked for Andrew's passport and told me not to worry about anything. Yes!

Then we heard that in order to honor the three-day Muslim feast of Eid-al-Fitr (the time of celebration at the end of the fasting month of Ramadan), the border with Syria was going to be opened temporarily in order to allow relatives to cross back and forth so they could celebrate the feast with their families.

When we heard that news, it was our cue! Like exiting a supermarket, we could see the door sliding open, and we jumped at the chance to make our move. We were on our way to the Lord's next appointment for us. We might have been a little naïve or even outright stupid, but Jesus didn't mind. He kept all the promises He had made to us, and we arrived safely in Greece after two days of driving. In northern Syria, we were crossing through what was known to be a heavy rebel area and had been pulled over at a roadside check-point. Suddenly, we heard automatic weapons fire. It seems a car on the other side of the road had not stopped as quickly as the police wanted, and they sprayed warning shots over the top of the car. They stopped fairly quickly after that. Except for a broken fan belt in central Turkey, the rest of the drive was uneventful.

Our plan was to find a cargo ship that could take us from Piraeus in Greece to Buenos Aires, Argentina. We had been communicating with some friends who lived near Athens about our intentions to come to Greece in order to find a freighter ship that would take us to South America. They graciously invited us to park our caravan in front of their house, but with regard to finding a ship that would take us to Argentina, they were not so enthusiastic about our chances. They told us that they knew of some other folks who had recently tried to hop a freighter ship in that direction, but to their disappointment, they had moved on to Italy when they were not able to find a single shipping line serving South America.

This was, of course, not the kind of news we had hoped to hear, but we had faith that if it had truly been by divine appointment that we had received our instructions, then He would have to provide for us according to His promise in Philippians 4:19: *"But my God shall supply all your needs, according to His riches in glory by Christ Jesus."* It was this actual living out the Word that was making it come alive for us. God likes us to step out by faith and claim His promises. He enjoys watching as we move out onto the branch and begin sawing it off, expecting that He will catch us and support us. And HE DID!

After just a few days of parking outside the house of our friends, we were sitting together during dinner when one of them shared the most amazing news. "You'll never believe what happened to me today!" he said with great excitement. "I was hitchhiking home, and a man picked me up. We began to chat, and during the conversation, I asked him what he did for a living. He told me he was the First Officer on a freighter ship. I said, 'Oh really? Where is your ship going from here?' And he calmly said 'Buenos Aires, Argentina.' Trying to contain myself, I asked him before getting out of his car if he could tell me the name of their company's agent in Piraeus. Here it is!" This was such an amazing answer to our prayers that we *knew* it was God!

The next morning we hurried down to the crowded and confusing harbor of Pireaus, one of the busiest ports in the world, to find the agency for "our" ship. You can imagine how surprised the receptionist was when she asked how she could help us, and we replied "We would like to take our car and caravan and travel on your ship to Argentina!" Taking a moment to recover from the shock of this statement, the poor lady went off to consult with someone in the back office. A man returned to meet with us, and we took time to explain that we, along with our two young children, were on our way with our car and caravan to do mission work in Argentina, and that we wanted to hitch a ride on the vessel that was currently in port. Amazingly, he was sympathetic enough but explained to us that it was not all that easy. The good news was that it was not entirely impossible either. The bad news was that this particular vessel would depart in just three days, and there would not be enough time to contact the head office of the company in Switzerland to get the required permissions. Not disheartened by this reply, we asked if they could at least send a telex to the head office with our request (this was in the pre-internet days), which he agreed to. There must have been a soft spot in the hearts of those office workers to inspire them to try and help us, since what we were asking was crazy to say the least.

God's plan is often like a puzzle that has many different pieces that all need to fall into place before the full picture can be seen. During those days, a dear friend of ours whom we knew from our days in Malta was passing through Athens on his way back to Libya. He worked in the Libyan oil fields and had been enjoying his once-every-two-months time off from the desert. We met up that evening, and during the conversation, the subject of our journey to Argentina and the ship we had found came up. We told him that we were encouraged by the response of the agents but that there was probably not enough time to get a positive response from the head office in Zurich before the ship was scheduled to depart in three days' time. He said, "Zurich? My flight to Tripoli tomorrow goes first to Zurich where I will change planes! Why don't you write up a request letter, and I will drop it into the post at the Zurich airport?"

Now, who planned that route? Suddenly, we could see all the heavenly bodies lining up in perfect conjunction, working together to make this miracle a reality. We took that evening to write up our official request, explaining our need and intentions, and asking if they could help us with free passage on the ship currently in port. We prayed over the envelope, gave it to our friend, and went to sleep confident that it was now all in God's hands.

The next day we received a call from our friend to tell us he had successfully posted our letter from the Zurich airport and had sent it by express special delivery to get it to the owner's office as quickly as possible. We hurried down to the shipping agent to see if there was any news, and the man with a smiling face said, "I don't know how you did it, but the owner was so impressed with how you managed to get your request to him so quickly (it had arrived not long after the telex from the agent), that he has agreed to your request for passage! There is not enough time to finish all the details before this ship leaves, but there will be another ship passing through in four weeks, and you can take that one. You will need to pay something for your food, but you will stay in the owner's cabin, and your car and caravan will be shipped for free. You will also need to take out traveler's insurance, especially for the children. A ship can be a dangerous place!" We breathed a sigh of relief and praised the Lord for His providence and miracle-working power. We would use the extra month to prepare and begin learning Spanish.

CHAPTER THREE

NO TRIUMPH WITHOUT A TRIAL

Even though we had not been living by faith all that long, one valuable lesson of life I had learned was that often, before a milestone event, there is a test. "No triumph without a trial."

In this case, the test for me came in the form of a hospital stay and a life-threatening operation. For Andrew, a debilitating sickness stopped him in his tracks.

I had always been quite healthy and had not had any complications or difficulty bearing our two children.

Andrew, Anne, Joy, and Aaron

Anne with Aaron

We were not practicing any form of birth control and were trusting the Lord for how many children we would have—something that in these more "modern" days might even be considered "irresponsible." For us, committing the number of children we would have to the Lord was a non-issue.

I had been experiencing some abnormal bleeding since my last period. We mentioned it to our friends who told us they could refer us to a gynecologist they knew. I decided to give it some time, and we opted instead to take a trip outside of Athens for a while where I could rest, and we could experience the Greek countryside. We had remained mostly in Athens up until that time. After all, we had a month to wait before our ship, the Penelope II, would arrive in port.

We had not driven very far and were taking a walk in a village about forty kilometers outside of Athens when I felt I needed to sit down on a park bench. After a minute or two, I felt faint, and Andrew had to support me as I collapsed into his arms. We immediately called our friends, who gave us directions to the hospital where the gynecologist worked. They also called him to let him know we were coming in.

I was admitted to the hospital, and they began administering tests, but, for some reason, they could not identify the problem. All they could do was pump me full of antibiotics and hope that somehow that would help. In the meantime, Andrew had parked our caravan alongside the soccer field of an elementary school that was adjacent to the hospital. During the day, he would take care of Joy and Aaron, but there was no electricity, and it was very hot. In the evening, they would drive to the house of our friends to have dinner together and so their children could play with ours.

Andrew recounts:

"On the third day of Anne's hospital stay, while I was driving home rather late into the night, I began to feel feverish. I knew I was coming down with something, and for the last fifteen minutes or so of the drive, I was praying desperately that I would be able to handle the car so we would reach "home" safely. Thank God, the children were well behaved, for the next day I was deathly sick and had the tell-tale sign of yellowed eyes, revealing that I had indeed contracted hepatitis! I could only conclude that it must have come from a drink I had taken from the garden hose during the time we were parked in front of our friends' house.

How I managed to endure caring for the children during those days, I do not know. Joy did such a good job of entertaining little brother Aaron. The clinic asked me for a urine sample, as urine turns dark when there is hepatitis. When I handed my sample to the attendant, she asked 'What is this?' I said 'It's my urine sample.' It looked like Coca-Cola! The Lord was surely giving me both the grace and the faith to handle that situation.

They say that when gold passes through the refining fire, it comes out even purer gold. I could only cling to that hope as I heard from Anne a few days later that there was bad news."

After spending six days in the hospital and not finding out what was wrong with me, the doctor said they were going to release me. I asked Andrew to pick me up the next day. Then that morning, a visiting doctor who had heard about my case came and did an ultrasound and discovered right away what was wrong. She could clearly see that I had an ectopic pregnancy. A fertilized ovum had lodged itself in the fallopian tube rather than the womb. It had grown for a time but then died, and the bleeding and eventual pain was caused by the resulting infection. We could not understand how the other doctors did not determined that earlier, but at least we now knew what we were facing.

When Andrew arrived, thinking he would be picking me up, I had to tell him, through my tears, that they would perform the operation that afternoon that would remove one side of my tubes and ovaries—something that all but eliminates the possibility of future pregnancies. Andrew and our friends all donated blood and prayed that I would be okay. The devil was trying to take not only my life but the life of any future child I might bear, but the Lord was in control of every detail. He had a plan for us that the enemy could not stop, only hinder temporarily.

There was one last hurdle in front of us—the hospital bill! As I was about to be discharged, I slowly made my way through the foyer of the hospital to the cashier's window, and asked the amount of the bill. Not having the money, I explained why we were in Greece and that we would soon be going to Argentina as missionaries. The angelic cashier said, "That's okay then. We'll take care of it. You are free to go!" Just like that, it was over. We had gone through the fire and passed the test! Even in the midst of our darkest trial, He was there taking care of us. He does all things well.

What if we had made it onto that first ship and then all of this had happened? Initially, not making that ship had seemed like a defeat, an unnecessary delay. But God's delays are not denials, and He always knows best. In the 1950s, there was a popular TV program called "Father Knows Best." That's definitely true. Our Heavenly Father, that is!

We were both thankful for the time we now had to recover and counted our blessing while waiting for our ship to come in. Then, finally, it was time! Penelope II had arrived in port. There were just a few details we needed to work out with the agency, and then we would be on our way at last. We were informed how much we would have to pay to cover our food and that the car and caravan would be loaded the next day at 8 a.m. Embarkation would be from 8 a.m. to 10 a.m., and the ship would depart at 11 a.m. sharp. We were ready. We had passed through the fire and come out victorious, or so we thought. But there was more.

For two years, we had been carrying with us a 1,000 Swiss franc note (worth approx. $400 at that time), a little "survival" fund that was only to be used in case of an emergency. Now we needed to exchange it in order to pay the agency. No problem, right? There were a multitude of banks interspersed among the shipping agents in Piraeus. We entered into one and sauntered up to the foreign exchange counter, handing the note to the teller. He looked at it curiously for a moment and began to sift through a ring binder that apparently had photographs of bank notes from every country on earth. When he reached the pages for Switzerland, we could see the photo of our note, yes, it was genuine. "But," the teller said, "Our bank will not exchange this note." He said something about it no longer being in circulation. We were not, however, about to let our plans be spoiled by the devil. We walked confidently into the next bank, and, this time, the exchange was done quickly and expediently. "Another angel?" we asked ourselves as we returned to the agency, money in hand and smiles on our faces. Either way, God was in control!

Since the following morning's embarkation was scheduled fairly early, and we were not sure how congested the traffic would be as thousands of commuters would all be trying to get to their offices at the same time, we decided to drive our car and caravan down to the port area during the latter part of the evening and sleep there. That way, all we would have to do would be to wake up, dress the kids, and drive on to the loading dock a few blocks away. Piece of cake, right? Wrong! The drive down to the port area was easy enough, as was finding a suitable place to park our elongated rig. We found a nice, big, totally empty parking lot, took a convenient place right in the middle under a street light for additional security, tucked the kids in, and quickly fell into a contented sleep. The worst was now behind us, and from here on in, it would be free sailing ahead, or so we thought!

We had set our alarm to make sure we were up bright and early for the big day before us, but before the alarm had a chance to go off, we were awakened by voices and commotion outside. Pulling back the curtain and peeking out to see what was happening, we were shocked to see that our vehicles were now totally surrounded by a sea of parked cars. What we had not realized when parking the night before was that the Greeks all arrive and depart at the same time, and because there is a shortage of parking, no one parks in the designated spaces. The cars had parked bumper to bumper, blocking us in by at least seven cars deep.

We stepped outside to assess the situation, and it did not look good. Even if one or two of the cars had been left unlocked or in neutral so they could be pushed, that would not help, as the entire lot was in grid-lock. Nothing could be done until all the drivers were there at the same time. The thought passed through our minds that we could actually miss our boat. So close, and yet so far away!

As we walked from car to car, looking inside to see if anyone had left their keys and trying doors to see if any cars were open, we began to attract attention. Once we explained our dilemma, a few locals stepped up to help us. They began honking their horns loudly to get the attention of the office workers in the surrounding buildings. Soon people were looking out of their windows to see what was going on. Apparently, the visual image below was clear enough that before long, the owners of the cars that were blocking us in began to appear, and within about fifteen minutes a path had been cleared so we could pull out.

We felt like Moses and the children of Israel passing through the midst of the Red Sea, and to be honest, it was a miracle on a comparable scale! Smiling Greek office workers waved to us as we headed out of "Egypt" on our way to the Promised Land.

CHAPTER FOUR

SUPERNATURALLY SAVED AT SEA

There was an air of excitement as we boarded Penelope II. It was our first time to cross the Atlantic and our first time on a freighter ship. We watched as the crane lifted first our car and then our caravan onto the deck. Andrew questioned one of the Greek officers as to why they were placing the caravan aft, on the leeward side (the back left corner). He explained that, in a storm, that position would provide the greatest amount of safety to the caravan. That decision turned out to be prophetic.

Vehicles ready to be loaded on board

Waiting to go on board

We were assigned to the owner's cabin, which is reserved for the owner or special guests, and ate each meal at the captain's table with the officers. We learned much about sailing and navigation, and although the children were young, they were fascinated with it all. We pretty much had free reign of the ship and could visit the bridge or map room whenever we liked. Only the engine room was off limits without permission.

As we got underway, the captain explained to us that we would first be going north through the Dardanelles and the Bosporus Strait to the Black Sea, where we would then take on additional cargo at the port of Constanta in Romania. After that, we would set sail and not stop again until we reached Buenos Aires.

In Constanta, we were able to go ashore, see the sights, and shop if we liked. This was during communist times, and there was a thriving black market in the port. The sailors on our ship would change their dollars with the dock workers in order to get a better exchange rate than the official banks would offer. So we asked them to take care of that for us. Each visitor to the port had a "cigarette quota" that enabled them to buy a limited number of cartons of American cigarettes (which were rare and extremely valuable on the black market). The sailors who did not smoke would buy their quota of cigarettes anyway and then sell them on the black market. When exiting the port area we would often be approached by shady-looking characters asking us if we could buy cigarettes for them. We left all that up to the more experienced crew on our ship. We did not need to buy much, since all that we needed was being supplied on board. Besides, if you have never seen a department store during communist days, you haven't missed much. Picture big, rundown, and grey on the outside, with unfriendly employees and empty shelves on the inside. But we were thrilled with every new experience. On ship, we continued to improve our Spanish by talking with the crew members, who were primarily Argentinean and Uruguayan.

Each day, we would look out of our little stateroom porthole to see if the cranes were loading the cargo. To our amazement, it seemed they were more often idle than in use. We asked what was going on and why so little work was getting done day after day. It seemed to be a combination of uninspired communist workers (who got paid regardless of how much work they did) and a bureaucratic problem between the shipping company and the port. It looked like our three-week transatlantic cruise was going to take a bit longer than we had planned. But we did not mind the delay, since I was still recovering from my operation and did not have to worry about doing any cooking or housework while on board ship. Keeping the children occupied was our only responsibility.

The weather remained favorable, but winter was quickly approaching. We were familiar with the story of Paul's shipwreck in Acts 27. He had warned the centurion who was considering leaving the Fair Havens, *"I perceive that this voyage will be with hurt and much damage, not only of the lading and ship, but also of our lives"* (v.10). They, of course, ended up shipwrecked in Malta.

We prayed our fate would not be the same as theirs, or worse! But here we were, getting close to that season and about to pass by that very way. Still, the work dragged on slowly day after day until winter was fully upon us. The snowfall on the ship's deck provided wonderful playing opportunities for the children, and we were safe and warm inside, using the time to teach them things about the ship. We had a special home-schooling curriculum, and, at the same time, they were getting a lot of extracurricular knowledge from actually living their geography and sciences classes!

We spent a memorable Christmas with the captain and crew of our ship. In January, we endured a bitter storm that unfortunately resulted in the sinking of several ships that were at anchor out in the bay. Thankfully, we were safely moored in port.

The loading of our vessel dragged on into February, further complicated by the very cold weather until, finally, after nearly two months in port, we heard word that we would soon be setting sail. Oh, how we longed to pass through the Straits of Gibraltar and start heading south toward the sun and the summer season. But, first, we needed to pass through the Mediterranean Sea during the height of winter.

Heading south past Istanbul and down into the Aegean Sea below Greece was no problem, but as soon as we got out into the Ionian Sea between Greece and Sicily, we were hit by a winter storm with force-12 gale winds (one degree below hurricane strength). The captain assured us that ours was an older, strongly-built vessel and that we were heavily laden, which keeps the ship low in the water, so there was no real danger of capsizing. The ships that had sunk in port were empty at anchor, and just bobbed like corks until they fell over. Thank God, our trust was in Him, and not in circumstances.

We did not see much of the crew. The reeling to and fro caused even the seasoned sailors on ship to lose their lunch, and most had retreated to their bunks with seasickness. Andrew, too, was confined to his bed, but Joy and Aaron thought it was great fun and were having a wonderful time as we did our best to keep our composure and our stomachs where they belonged.

"Could this vessel actually sink?" we wondered. "Would this be the end?" Unlike Jonah, who was headed out of God's Will, we knew the Lord had shown us to go this direction and had supplied for us this very ship. So I cried out to the Lord for His help! "Jesus, please help us! Help me to get over the seasickness. And calm this sea!"

I stood up, and He gave me the strength to walk right up to the bridge, which was no easy task considering how the ship sways even more severely from side-to-side the higher you are above water level. We had spent many long hours on the bridge under calmer conditions—chatting with the captain, studying the navigator's maps, and admiring the majesty of God's creation. Now, it seemed that same might was set to destroy us! Upon my arrival, I found the captain there alone, guiding us through the storm. I joined him in his vigil. What a wonderful parallel of how our Heavenly Father, the Captain of our lives, sees us through every adversity. We altered our course and diverted below Sicily to find refuge from the blast of the wintry billows, and plied slowly along until the storm ceased. Some material damage was sustained by the ship's cargo, but no one was hurt. Like Paul, our prayers had been answered.

As we neared Gibraltar, the sun emerged, and the temperatures warmed considerably. Andrew went down to see if there had been any damage to our vehicles. A wave had apparently crashed across the car so that the roof was dented in, but he was able to push it back up, and it popped back into place. And, yes, the caravan was in perfect condition with no damage at all, having been placed in what the officers knew to be the safest part of the deck, and having been securely tied down with heavy cables.

We have no guarantees in life that we will never have troubles, but it is important to face the storms of life unafraid! Just climb up on top of the situation. Rise above it! Go up to the bridge and take a firm hold of the Captain's hand. Jesus is our Pilot, and together, we will enter into safer waters! It takes faith to pray, yes. But with guts and gumption to get up and act, the results will be wonderful! The everyday battles of life may sometimes seem overwhelming, but the victory is there, waiting right around the corner! And it is worth fighting for! Don't take the storms of life lying down!

Most people's natural tendency is to resist difficulty. Hide from it. Close their eyes. Hope it will go away. But why does a mountain climber seek out Nepal? Or a soldier request duty on the front lines? It's to face the challenge boldly and bravely, unafraid. They are seasoned veterans who relish the battle. Let's admit it, life is a battle, and we will be facing them until the day we die! Let's learn to face adversity, trusting our Pilot, and we will come through on the other side victorious!

And, so did we come through this trial unscathed. After that, it was smooth sailing all the way to Argentina! Crossing the equator included a ceremony by the crew that had been especially adapted for first-timers. We saw dolphins dancing in the sea and flying fish performing for us—more evidences of God's wonderful creation.

CHAPTER FIVE

Life In Argentina

Once we were into the open Atlantic, things began to warm up quickly, and we were able to finally begin enjoying the vacation side of our ocean cruise. Andrew thought it would be a good idea to go down and check out the vehicles—fix a bit of rust, clean, wax, etc., but when he checked the car's radiator, he noticed that there was no water in it, which was odd. He filled up a plastic bottle with water, but when he began to pour it into the radiator, he heard a strange gurgling sound of water dripping to the deck. Looking underneath he could see that most of the water he had just poured into the radiator was now on the floor. Straining to see where the water was leaking from, he discovered a crack about four centimeters long right in the side of the engine block. The winter temperatures in Romania had gotten so low that our radiator had frozen and cracked the block. What would we do now? As soon as we arrived in Buenos Aires, we were expected to hitch up our caravan and drive away. That was obviously not going to happen. We were going to need another miracle!

We had so much fun fellowshipping with the Latin crew members on board Penelope II. Most of them could speak some English, since sailors are an international lot. Having a young couple with two small children on board was a real novelty for a freighter ship, so they were more than willing to spend time with us, playing with the children, and helping us with our Spanish.

One of the boys whom we got to know quite well told us that he was making a very good salary in U.S. dollars, and even though it was a big sacrifice to have to live away from his family, he and his wife had agreed that he would work like that for five years, almost non-stop, until they had saved up enough money to build their own house, settle down, and lead a more normal at-home life. We admired his dedication.

The South American currencies were notorious for large devaluations, so earning their salaries in a hard currency like dollars was a big blessing for them. One boy told us he had bought a nice BMW in Uruguay for a fairly good price because he had exchanged his dollars into the local currency to make the purchase. A few weeks after he had bought his car, the government had devalued the local currency by fifty percent, meaning if he had waited to change his dollars until after the devaluation, he would have gotten the car for half the price he had paid! You just never know. A company that had purchased a fleet of buses promising to pay back the loan in U.S. dollars went bankrupt after the devaluation. Once we arrived in Argentina, we would have our own experiences in dealing with the hyperinflation.

The Rio de la Plata forms a large estuary as it empties into the Atlantic, and we could tell we were getting close to our destination when our ship passed from the aquamarine ocean into the muddy brown river water of the estuary. Before long, we could see the skyline of Buenos Aires before us. We were coming to the end of our three-month-long adventure!

Our Argentine contact person was waiting for us as the ship tied up in port. We introduced ourselves to him, but before we could get into any small talk, Andrew explained to him our predicament with the cracked engine block. Unperturbed, he escorted Andrew outside the port area to look for a solution, while I waited with the children alongside our rig as it sat on the dock.

Chapter 5: Life in Argentina

Andrew continues:

"We walked a few blocks away from the port to an area that had service stations and auto parts shops. We entered one store, and after explaining our situation to the proprietor, he handed us a tube of something called Acero Liquido (Liquid Steel). I had never heard of such a thing, but we bought it and walked back to the port. Locating the crack (thank God there was only one), we worked in the silver colored paste as best we could and then prayed and waited for it to dry. The moment of truth came when it was time to start the car, but, of course, the battery was dead due to lack of use during the bitterly cold winter. With an abundance of vehicles coming and going inside the port, it was not hard to flag one of them to come help us. With the help of jumper cables, our faithful old Peugeot started just fine. We let it warm up to operating temperature, and it showed no signs of leaking from the crack. And would you believe, we would drive that car thousands of miles over the next three years while pulling the caravan behind us, and that Liquid Steel would hold the entire time!"

We had come to Argentina to have a mobile caravan ministry. Our Spanish quickly improved. We came to love the Latin people and their culture, and the children continued their real-life schooling experiences. We lived in the caravan full-time, summer and winter. Since the country was so large, we could migrate to where the temperatures and weather were most agreeable. We were able to enjoy summers in the beautiful resort town of Mar del Plata, winters in the warmer sub-tropical Cordoba and Santa Fe, and even drive up to Mendoza near the Chilean border in the high Andes.

When it was time for our visa to expire, we would cross over the border to Uruguay, and continue our ministry there in Montevideo, the high-classed resort town of Punta del Este, and even up to Rivera on the border of Brazil. There you could sit in a café and have coffee with one person sitting in Uruguay and the other in Brazil, since the border went right through the town!

Following my operation in Greece, the doctors had told me that I might not be able to have any more children, but, lo and behold, during the voyage, I realized that I was indeed pregnant. Another miracle! I was about eight months along when we attended a three-day Christian conference in Buenos Aires. Since I did not yet have a hospital picked out where I would have the baby, they made an announcement at the conference that we needed both a hospital where I could deliver the baby, and a safe place in the city where we could park our caravan while we waited for the time to come. We were delighted when one of the Christian bands that had been performing (their music was all original and straight from Heaven!) stepped up and said we could park alongside their house, and that they were not that far from a hospital where they had also had babies delivered. The Lord was supplying. We had such an amazing time staying with these dear brethren, sharing testimonies, and listening to their amazing music.

I had a dream that I would have a little girl and that her name was Angelina. After that, I was listening to one of the band's songs titled "Angeles de Dios" (Angels of God), which was talking about when babies come down from Heaven. That was a confirmation, and I was sure it was going to be a girl. When the time came for me to go to the hospital, I was in the delivery room while Andrew was just outside in the hallway. One of the nurses attending the delivery was filling out some forms and asked me what the name of the baby would be if it was a girl, and I answered "Angelina." When she asked, "And if it's a boy?" I said, "It's going to be Angelina." She insisted that something needed to be written down in case it was a boy, and it could always be changed. She then went out to ask Andrew, but he answered the same. When he heard me huffing and puffing in the delivery room, he said "Just give it a few minutes, and you'll see." Sure enough, just a few minutes later, the nurse shouted out, "Yes, it's Angelina!" Because of my condition and having had the operation just months before getting pregnant, we knew that this was a special child sent from God—Angelina! Our little angel.

On the road in Argentina

When Angelina was only five months old, Argentina invaded the Malvinas Islands that were not far off the coast of Patagonia, reclaiming them from the British who had been colonizing them since the 1770s and calling them the Falkland Islands. It was a long-running dispute that had gone on for over two centuries, but when the British fleet set sail from Portsmouth and started heading south, we knew there was going to be trouble. We moved quickly to neighboring Uruguay. We found an Argentine circus camped on the outskirts of Montevideo, and asked if we could camp with them. Together, we would listen to the news and follow the events of the war, which was more often bad news for our friends.

It was not a good time to be English-speaking in Argentina. We had heard of English-speaking foreigners being pulled off of buses and beaten. The military junta was trying to secure and legitimize its hold on power through this war, following much international criticism as a result of their so-called "Dirty War" and the many "disappearances" during that time of government dissidents and left-wing extremists. One of our English friends who had only been in Argentina for a few months had to go into hiding until she was finally able to escape and get a flight back to the U.K. Also, during those days, the Argentine peso was hyper-inflating. Prices would go up daily on almost everything. The currency was changed, and new bills were issued after having taken four zeros off, so that a 10,000 peso note was now one new peso. Times were crazy.

When we felt it was safe to return, we were driving through the area of no-man's land between the two borders and stopped briefly to prepare our passports and car documents before crossing into Argentina. For some reason, we could not find the car papers where we normally kept them and could not figure where they could possibly be. As is so often the case when something is missing, we began to look through our bags, pockets, under the car seats, etc. They were nowhere to be found. Our anxiety started to turn to panic as our search moved to the caravan cupboards, under the bed, places where we would have never put them. In the end, we had to finally declare them as missing. In our search, we did find an old insurance policy from back when we had first purchased the car in the south of France. It had all the pertinent information on it, make, model, year, and license plate number, and it was written in French. So with a prayer and a look to Heaven, we proceeded to drive up to the border. We handed our passports and car documents to the official on duty, and he stamped us in and waved us through without a single question! We never did find the missing car papers.

After three very fruitful years, we again felt the call to change fields. This time, we would travel to the Indian subcontinent, specifically Bangladesh. Since we enjoyed so much our first transatlantic cruise, we wanted to again find a freighter ship that could take us to our new field, either directly to India, or possibly Singapore, but we could not find anything. At the same time, my father was inviting us to come to Finland for a visit, as he wanted to see the children. After promising that if we would pay him a visit, he would pay our ongoing fares to Bangladesh, we felt like that was how the Lord wanted to supply for us this time, so we agreed, and began to look for a ship to Europe.

We would not be needing our vehicles any longer, so we decided to sell the caravan in order to raise funds for other needs.

Our old car was so run down and on its last leg that we did not even think of trying to sell it. Plus, we would need it to drive to the port of Buenos Aires to look for a ship.

We had bought the caravan in Jordan, so it had license plates written in Arabic. It had been owned by a Syrian family who had taken it to Mecca to perform the Hajj (pilgrimage) a few years previously. It was an older German model, fairly heavy, and so we expected it would not be so easy to sell! We parked by the house of some friends and put an ad in the second-hand newspaper, but no one seemed to be interested.

One day, we were joining our friends in a day of fasting and prayer, since they were having problems, and we needed to desperately sell our equipment. That day *two* buyers came at the very same time to see the caravan. They said they had seen the ad that day in the newspaper, but we had not put the ad in for that week. Then all of a sudden, both potential buyers started to argue over who would get the caravan, and the price they were offering started to go up! It was like an auction. While one was offering to pay us by installments in the quickly hyper-inflating Argentinean currency, the other took us discreetly aside and offered us an even higher price and payment in U.S. dollars. We shook hands and accepted his offer. A few hours later, they returned with the cash. We hitched her up to their car, and away she went. God had intervened in a big way, and the caravan was now sold for even more than we were asking!

The Port Authority published a daily paper of all the port movement—ships coming and going, along with their origins and destinations. We did not see anything suitable listed, but went to the port anyway to take a look for ourselves. We asked a dock worker if he knew of any ships heading toward Europe. He pointed out one that was named Copenhagen Maru. The connection to Scandinavia was obvious, so we went up on board to see if we could meet the captain and explain to him our need. Really, it was a miracle in those days how we could just drive into ports without any control, and walk up onto ships.

Copenhagen Maru was South Korean. The captain was kind, but did not speak much English. He told us that we would need the permission of his agent in the city. The agents told us that if it was okay with the captain, it was okay with them. We had a green light for our second transatlantic cruise. This time, we didn't even have to pay for our food. It was completely free!

Returning to Europe on Copenhagen Maru

The final miracle came when the sweet brother who had met us at the dock when we first arrived and had seen us going all over Argentina and Uruguay for the past three years in our faithful old car insisted on buying the car from us and offered us $200 for it, the exact same price we had paid for it five years and three continents before. We gave him our dog as a bonus gift, and he drove us to the port to see us off.

CHAPTER SIX

TO FINLAND AND BANGLADESH

We were happy and excited to be back at sea and on our way to new adventures, this time with a South Korean crew. It took us a while to get used to the new cuisine, which the chief cook would, of course, prepare to cater to the taste of his crew and not to us. They were kind enough to offer us hamburgers on evenings when the Korean food might have been especially difficult for us to handle, but for the most part we tried to *"eat such things as are set before us,"* giving thanks (Luke 10:8).

It was however on the third day at sea that I began to feel quite sick to my stomach. At first I thought it might have been because I was unaccustomed to the food, but when on the fourth day I awoke with tell tale yellow eyes, it became apparent that it was now my turn to have hepatitis. During our last few weeks in Argentina there had been severe flooding of both the Rio Parana and Rio Uruguay that form the Mesopotamia region of the province of Entre Rios, just north of Buenos Aires. The media had been warning people not to drink the tap water, but apparently I had taken in some tainted water somewhere and had contracted hepatitis while at sea. Because we had done quite a lot of research into treating hepatitis when Andrew had it in Greece, we knew that in addition to rest, a non-oily, high protein diet was of the utmost importance so the liver could recover without subjecting it to too much work. The problem was, how would I be able to manage such a diet on board ship where we were pretty much forced to eat what was being prepared for the rest of the crew?

Because Copenhagen Maru was heading all the way to Arhus, Denmark, we needed to stop in Recife, Brazil, to load additional fuel before heading on to the transatlantic part of our voyage. We would only be in port for about six hours, but that was enough time for Andrew to go ashore and search out some beef liver for me, one of the highest protein meats around. Counting at least fourteen days we would be at sea, he bought an entire liver that weighed about four kilograms, the largest one he could find. We did not want to trouble or worry our South Korean hosts, so we did not explain in detail to them how sick I was. When Andrew brought the liver on board and asked the cook to put it in the freezer, they must have figured we were tired of their home made cuisine and shook their heads each evening when we asked them to boil up a portion of the liver for me, since I was not able to eat anything fried.

The "crossing the equator" ceremony was a bit more exotic this time with our Asian friends. It included candles, an assortment of trinkets, and a pig's head. We were respectful the whole time but were happy when it was over. Other than that, there was not a whole lot of excitement until we reached the English Channel between the cliffs of Dover and the French coast. It was late at night, and outside our portal window, we could see red flares being fired across our bow. Curious what it was about, we headed up to the bridge. There, in the middle of the busiest waterway in the world, our captain was nowhere to be seen, while our ship was drifting outside the designated transport lane. The channel police had been trying to contact the bridge by radio, while firing the flares to get the ship's attention. The officer on duty could not understand the English accent as it was coming across the radio, so when we appeared on the bridge, he asked us if we could help interpret. We stepped in to save the day and took up the radio to communicate with the police boats on both sides of us. They would give us the bearings to turn to that would guide us back into the channel, and we would then pass them on to the first officer. We were thankful that we were able to help avert a possible tragedy.

As we were nearing the port of Arhus on the west coast of Denmark, it seemed like a welcoming committee of ladybugs were sent out to greet us, as when we were still quite far out, the entire deck was completely covered by the friendly red beetles, to the delight of our three children.

Moving back to a frigid winter climate from South America was a thrill for the kids, but for us, it took some getting used to! We traveled overland by train up to northern Sweden on our way to visit my father in my home town in northern Finland. As we crossed the border the temperature was dipping to around -25 C.

On a subsequent visit by train to northern Finland

We enjoyed our two-month stay with my father and visiting all of the relatives whom I had not seen for many years. But the time came for our visit to end. We needed to continue on to our next field of Bangladesh!

In giving his invitation to come visit, my father promised that he would pay for our ongoing flight. His actual plan was apparently to try to strand us in Finland with no money or way of escape in the hopes that we would then settle down and stop gallivanting around the world! We had been giving our three small children an excellent education through our home schooling methods, but many Finnish people did not understand how it could be possible to educate children outside the established educational system. It was obviously time for another miracle, and it came after just a few days!

One evening I found a curious looking envelope addressed to me lying on the dining room table. Opening it, I read that the local government wanted to buy a small strip of land that was in my name, although I had absolutely no idea at that point that this piece of land belonged to me. They needed the land to put a bicycle path alongside the road on which the land bordered, and they needed my signature to finalize the deal. My father, having realized that his plan had been thwarted and that the money from the sale of the land would be enough to pay for our plane tickets to Bangladesh, had begun drinking and left the letter laying out where I could see it!

We did not have any fixed support from churches or the government, but we served a great God Who had promised to supply our every need, even sometimes in unconventional ways. We never lacked because money was not the object of our existence. Making the world a better place was our goal, and educating and training our children in His ways. Book knowledge is one thing, but we wanted to give our children something we had never learned when growing up—real life experiences. That was very important to us!

Our flight to Bangladesh was on Russian Aeroflot Airlines, and we had a stopover and change of planes in Tashkent, Kazakhstan. It turned out to be very interesting, indeed, when for some unexplained reason our ongoing flight was delayed. Not understanding the language, we had to just sit there trusting that, sooner or later, our flight would continue on. Finally, it did, and we were again on our way.

As our plane prepared to land at the airport in Dhaka, we looked out the window as the beautiful lush jungle landscape gave way to poor-looking housing developments. The apartment buildings were discolored black due, no doubt, to the mildew that grew so fast in the hot, humid conditions. We knew Bangladesh was one of the poorest countries on earth, and we were prepared for just about anything. What we weren't prepared for was when the people who had come to meet us showed up not in a taxi but in a horse-drawn cart!

Being situated geographically at the top of the Bay of Bengal was not an advantage for this poor nation, as it was very prone to extreme weather conditions such as cyclones, tornadoes, and large-sized hailstones. Flooding was also common, since the country is comprised of delta land from the Ganges River and is generally just a few meters above sea level. Our three young blond-haired children were quite the attraction when we went about the market areas. Once, we were in a shop looking around, and when we turned to leave, we saw at least twelve people staring at us through the glass like we were monkeys in a cage at the zoo.

During our time there, we lived in three different cities, including Chittagong to the south where we dedicated ourselves to helping the poor in whatever way we could. Naturally, in such an impoverished land, especially in the tropics, there is a lot of sickness and disease. Thankfully, we were pretty much spared from contracting any serious sicknesses thanks to God's protection as well as the strict health and hygiene guidelines that we kept. We did get amoebic dysentery once when we were visiting some neighbors. They offered us a sweet nectar from the date palm tree, and, out of respect, we drank it. Sure enough, a few days later, we all had sick stomachs due to amoebas! There was an expensive, extremely strong treatment available that we had heard had serious side effects on those who take it, but we preferred to use the local cure. As is so common in most areas, there are natural cures for the local diseases. We were told that fasting for three full days and eating only natural, unsweetened yogurt, along with garlic, would totally get rid of the amoebas. And it did! We all felt cured, but needed to check. One big business in these poor countries is clinical laboratories that analyze stools to ascertain the cause of your sickness. That was how we knew it was amoebas in the first place. Well, lo and behold, when we got the laboratory results back, they confirmed that we were totally clear! One of my former jobs had been in pathology, so I always leaned toward laboratory tests when facing any sickness! We were thankful for natural healing, as medicines were often made locally and in unhygienic conditions.

CHAPTER SEVEN

Bangladesh, India, and Pakistan

It is difficult to describe life in the Indian subcontinent with three young children. On the one hand, it's amazing that we actually had the faith to go to such an area known for its radically different culture than anything we had ever experienced before, extreme poverty, sickness everywhere, etc. But on the other hand, what a wonderful opportunity it was for us to share the love of Jesus with the people there, while at the same time, enjoying the myriad experiences and getting to know the people and all they had to offer. The children had great fun as they were often the center of attention. We had entered into another chapter of their real-life studies of the world, and they absorbed it all like little sponges—all the rich sights, sounds, tastes, and smells of the region. Now that they were getting older, we were all the more thankful for their correspondence school curriculum.

Making trips to the border when it was time to renew our visas was not as easy as it had been in South America. The distances were greater, and we no longer had our own vehicle, so we had to depend on public transportation, buses, and trains. We could get a three-month extension to our stay rather easily, but then after six months, we would have to leave the country. Some people we had met who lived in Calcutta had invited us to visit them if we were ever in the area, so we decided to make our first border trip there, one of the best known and poorest cities in the world. On the map, or "as the crow flies," it seemed like it would not be that big of a trip, but traveling overland through the marshy delta on poorly maintained roads and then crossing the mighty Ganges River turned into one of our first great adventures.

The first bus took us from the capital of Dhaka on fairly good roads, and things went fine to that point. What we had not been told was that the road stopped at the river. We would then have to take a "ferry" across the river and board another bus that would be waiting for us on the other side. We were on one of the typical long-distance buses that would have people's bags stacked high on the roof. If there was a shortage of seats, you could see people sitting up there, too.

It was dark when the bus pulled to a stop. We were alarmed when we saw what appeared to be an army of "coolies" (the lower-caste porters who traditionally carry bags on their heads) climbing up onto our bus and handing down all the bags. Looking out the window, we could see our suitcases being carried off and disappearing down the dark path. In those sorts of situations, you generally learn by observing, as we watched the other passengers exiting the bus and joining the procession. It was not so easy since the children had fallen asleep. We got off as quickly as we could and anxiously pursued our bags, hoping that they would reach their destination safely and not wander off into the night to some poor person's abode. But we had no reason for alarm, for there they were, neatly stacked on the ferry, our coolie standing at attention waiting for his tip.

The ferry was an old, wooden, flat-bottom boat with no electric lights. We made our way down into the sitting hold where places on the hard wooden benches were at a premium as we were among the last to arrive. In the middle of the hold was a single kerosene lamp hanging from the rafters and swinging gently side-to-side, its yellow glow not quite illuminating the entire room. Folks squeezed together to make room for us, and we tried to make ourselves comfortable as we again imagined the apostle Paul sailing toward Rome, staring at a lamp not unlike ours. This was the type of ferry that you would often read about sinking from overloading, but thankfully, our midnight crossing of the Ganges was without further incident. The ongoing bus was waiting for us on the other side, and the suitcase-carrying coolies were there again, patiently awaiting our arrival. By this time, we had learned the ropes, something we seemed to be doing continually, and thanked God when we could finally settle in to continue our trip to Calcutta.

During our two years in Bangladesh, we stayed primarily in the capital where we worked mostly with the student population. It was very easy to visit the campuses or enter one of the dormitories and strike up a conversation. As foreigners, we continued to be an attraction, and most people were happy for the opportunity to speak with us, hear why we were there, and practice their English. We did make several trips down to Chittagong in the south, where there was also a university. We were hoping to one day be able to visit the "hill tracks" toward the Burmese border, but that was a restricted area where you needed a special government pass to enter, so unfortunately, that never worked out.

Often in our travels we would stay in "guest houses," which were small, privately-run hotels, but without a lot of the more modern amenities that would be expected in a real hotel. They were inexpensive and convenient enough. For health reasons, we never ate out and cooked our meals in our rooms. To make sure the meat was thoroughly done, we pressure-cooked everything. The local people often cooked on a small pump-kerosene stove, something we had never seen before. It was made of brass, and on the kerosene tank base was a little stick that you would pump to build up pressure. Then when you opened the valve, the pressurized kerosene would hiss out in a vapor that you could then light like a camping gas stove. Once lit, it would still make the hissing noise and sometimes sounded like a steam locomotive: shush, shush,

Aaron tries his hand on a cycle rickshaw

shush. Some of the guest house owners did not like anyone cooking in the room, so we would have to be very discreet when using the stove, sometimes even cooking in the bathroom with the door closed, so the owner wouldn't know what we were up to.

In India, we spent much of our time visiting the private Catholic schools with educational materials. These schools, having been built by the British during the colonial days, were usually located in the higher elevation areas where the weather was cooler in the summer. This is where the British and the upper class Indians would send their children to get a higher-quality education than was available in the poor, state-run schools. We enjoyed our visits to these hill stations and actually spent two summers high in the Himalayas in Srinagar, capital of the Indian state of Kashmir.

This area had been under dispute for many years as both India and Pakistan had claims to it, and occasional border skirmishes were reported. We would often meet members of the U.N. peace-

Angelina gets a ride with our luggage

keeping contingent that were stationed there. Thankfully during the times we were there, things were quiet, and foreigners were able to visit. The area greatly needed the financial boost from tourism. Sadly, in the years after that, Srinagar suffered a lot from terrorist and guerilla activity that closed the doors to tourists. It remains a dangerous and contentious area to this day. We are thankful for the opportunity we had during those days to not only enjoy this very beautiful region, but to share God's love with the poor local people.

In Bombay (now Mumbai) and Goa, we would hold weekly "foreigner's meetings." India was a mecca for young travelers, many of whom were not your typical tourist but who were genuinely seeking spiritual enlightenment. So much has been made of the gurus and maharishis of the eastern religions that young people honestly searching for enlightenment and answers to life would go to India in hopes of finding something more. Their churches had left them cold, and drugs had left them unfulfilled. So, they thought that perhaps the answers might lie in India. Unfortunately, many would discover, much to their dismay, that Hinduism and eastern religions were not the answer. Only then, it would be too late, since they would find themselves out of money, out of hope, and often sick. We were able to explain to them how the truth is found in the Bible, and that Jesus is the Light they are searching for. Many returned to their countries with new vision for their lives.

While in India, we met people of all types: Hindu, Muslim, Sikh, Christian, and Jain, and we had learned enough about these different religions to be able to relate to the people fairly well. However, because of our extensive experience in the past with Muslims, we felt more "at home" when the time came for us to move on to Pakistan. Joy and Angelina loved to wear the local "Punjabi suit," and they fit in very well as they strove to "become one" with the local culture.

Pakistan was not at all like Bangladesh or India. It was much more highly developed, and there was not nearly as much visible poverty. Actually, we experienced a bit of reverse culture shock when we arrived in Karachi, the commercial capital on the Arabian Sea. We had become so used to slow, older model cars in India, the huge traffic jams, cows sitting in the roads, and ox carts plying along at a gentle pace that we were not prepared for the higher-paced life of Pakistan. We began a weekly meeting, similar to ones we had hosted in India, and we were hitchhiking to the hotel that had offered us a free conference room on Sundays. Andrew and I

Chapter 7: Bangladesh, India, and Pakistan 57

Joy and Angelina love to wear the local "Punjabi suit"

stepped off the curb and stepped half a meter into the road in order to flag a car when the light would turn green. Our Pakistani friend said, "No, don't stand in the road! It's very dangerous!" We did not understand why until the light changed, and all the newer model Japanese cars zoomed off as if it was the start of the Indy 500. But we enjoyed this change, and adapted quickly to life in a developed Muslim country. The southern portion of the country was more desert, but to the north it was greener. We were able to spend time in the beautiful cities of Lahore, Hyderabad, and also Abbottabad in the foothills of the Himalayas where they finally found Osama bin Laden some 25 years later.

While we were under no illusions as to how potentially dangerous living in Pakistan could be, it was not at all as bad as it is nowadays. I believe we were living in a bubble of God's heavenly protection during those days. The following harrowing story is proof of that.

It was during a visit to the inland city of Hyderabad, located about sixty kilometers north of Karachi. We had been there for a few days visiting local friends when one evening while having tea with the ladies, their husbands came home and burst into the house terrified and shouting. They informed us that there had been an incident in the city and that people were demonstrating and violence was breaking out. They said that we, as foreigners, were at particular risk, and that we should leave right away! None of the men, however, wanted to risk driving us back to the home of our host for fear that an angry crowd might try to burn their car! It wasn't easy, but we finally found a motor-rickshaw driver who was willing to risk his life for our sakes, and he drove us to where we were staying. We learned that the reason for the rioting was that the mayor had been shot and was feared killed, and his party was, of course, accusing the opposition.

In Pakistan, a lot of influence is wielded by tribal leaders and land owners, and we were staying with an important local land owner who happened to be a member of the opposition party, so he was especially at risk. But being a polite and hospitable Muslim who was worried about the safety of his guests more than his own well-being, he insisted on driving us to the long-distance bus station where we could catch the next bus back to Karachi. We all climbed into his white, four-wheel drive Jeep and headed toward the city! When we reached the bus station, we located the next bus that was due to depart for Karachi any minute, but before we could climb aboard, we saw several masked gunmen coming out of the bus. They had just finished robbing its passengers! We also saw snipers on several rooftops. The whole city was closing down, and shop owners had already pulled down and locked their shutters.

We had never been in a situation like that and should have been terrified, but it's amazing how the Lord can bring a spirit of peace over you during such times of imminent danger. We piled back into our friend's Jeep, and he made a quick exit from the bus station and headed toward another part of town where there was a smaller bus station that catered to "chicken buses," buses that the poor people took, often carrying large sacks of their produce along with a chicken or two. Thankfully, we got seats on the next "chicken bus" to Karachi and safely reached home. We were very thankful as always for the Lord's protection and His angelic guardians!

Another miraculous escape was reported to us by a Pakistani friend of ours who had only recently become a Christian. He had been driving to a certain market to get his hair cut when he was forced to stop at a red light. He waited patiently for the light to turn green, but it did not. After waiting a reasonable amount of time with the light remaining red, he was tempted to just drive through it, but he resisted the temptation. Then while he was sitting there, he suddenly saw in front of him a huge explosion. A bomb had gone off right in front of him in the market that he was heading toward. He knew it was the Lord who had kept the light on red, and thanked Jesus for saving his life!

CHAPTER EIGHT

POST-WAR YUGOSLAVIA

The fall of the Berlin Wall in November 1989 marked the end of one era and the beginning of another, an era of hope for the millions of people who had been trapped behind the "Iron Curtain" since the end of World War II. It was an era for us to begin a new ministry reaching the people of eastern Europe. Those who had been forcibly kept from the things of the west now had an insatiable thirst and desire to experience what they had been missing, especially the younger generation who had been born during the time of the wall. They wanted rock music, blue jeans, Coca-Cola, and American movies, but they also wanted to know about the things of the Spirit. They wanted to hear the gospel and get Bibles, lots and lots of Bibles! We wanted to play a part and make ourselves available to share God's love with these people who were thirsting for the water of life.

We made trips to Russia and lived for a while in Poland, bringing aid and the good news of Jesus. It was an exciting time. We took backpacks stuffed with gospel tracts and literature, and stood at the bottom of an escalator that was bringing people down into the subway. Once the commuters saw the commotion and what was going on, we would immediately be swamped by hordes of people reaching out in desperation and hope to get a copy of what we were handing out. We often had to stand on a chair to get above the crowd and then placed the tracts into the outstretched hands. We made sure there was a contact address printed on the literature so people could write in for more information, and write in they did, by the thousands!

Along with the changes in people's personal freedoms came big changes in the political structure of the former East Bloc countries. The Soviet Union broke up into its former republics, Czechoslovakia was divided into two, and war was raging in Yugoslavia. One day while watching the news on CNN, we saw a report of a mortar attack that had killed many innocent people in the central market of Sarajevo in Bosnia, and the Spirit spoke to us clearly: "You have to go there and help those people." For a long time, our hearts' desire had been not to just visit these former Communist countries, but to actually live there long-term, really investing our lives for the people. We heard that a couple we knew had gone down to Slovenia, the western-most republic of the former Yugoslavia, for a month-long exploratory visit. They were so overwhelmed by the suffering and need of the refugees there that they wanted to set up a permanent outpost and were looking for volunteers to help them. We were quick to reply. My father had recently passed away, and I had a small inheritance that would help with the cost of getting things set up.

Slovenia had been able to secede from Yugoslavia and was spared from serious fighting, thanks to the quick recognition of neighboring Italy, Austria, and big brother Germany. Therefore, it became a safe haven for refugees from the other Yugoslav republics that were suffering more in their struggle for independence, mostly Bosnia and Croatia.

The newly-formed Slovenian government did its best to manage the increasing refugee problem, and quickly set up centers in hotels, school dormitories, and wooden barracks constructed specifically for the emergency. We contacted the ministry responsible for refugees and obtained a list of the thirty or so refugee centers that had been set up all over the country. At the same time, we took trips into areas of Croatia that were bordering Slovenia, including the Istria peninsula near the Italian border where many refugees had also fled. We put together a children's program with music, clowns, and meaningful skits, and systematically visited every refugee center on our list.

Bosnian Hatidja helps in our show

Typical school program

These families had been uprooted overnight, and their lives were torn apart. The children did not understand and were often traumatized. We wanted to make them smile, and laugh, and feel like things were okay, that things would get better soon, and to also know that Jesus loves them. It didn't matter that many of these refugees were Muslims from Bosnia. They were so thankful for the love we shared and to know that someone cared, and they readily received us and our message.

Although our ministry with the refugee centers was extremely satisfying and fruitful, our ultimate goal was to go into central Bosnia where the worst of the civil war had been fought. In Slovenia we met a young Bosnian woman whose family had migrated to the capital Ljubljana when she was young. She volunteered as translator for us when we visited the camps, and we became very close. Her grandparents lived in the very northwest corner of Bosnia, in the countryside near Bihaça city that had been very badly affected by the fighting. She had not seen them since before the conflict began, so we planned a trip together to see things first hand and assess the situation, whether or not it would be safe for us to start making trips regularly into the heart of Bosnia. That trip was very eye-opening, and revealed the true amount of devastation that had been leveled on that area.

Damaged house in central Bosnia

At the same time we were also making trips farther into the south of Croatia and felt that until the situation in Bosnia had stabilized more, it would be better for us to make our next move into Croatia.

The Dalmatian coast of Croatia on the Adriatic Sea (*yes, it is where the dog gets its name*) was the summer playground of the former Yugoslavia. People from all over Europe flocked to the coast to enjoy the crystal clear aquamarine water, the pebble beaches, and the excellent wine and cuisine of the region. To the far south was the ancient walled city of Dubrovnik, which dates back to the 8th century and reached its height in the 15th and 16th centuries when its thalassocracy rivaled that of the Republic of Venice. In 1979, the city of Dubrovnik joined the UNESCO list of World Heritage Sites, but sadly, in 1991 during the height of the civil war, Dubrovnik and all the surrounding region endured serious shelling from the hills overlooking the coast and suffered much damage. One day while driving in Ljubljana, we saw a billboard advertising Dubrovnik. Obviously, Croatia was trying to reestablish its tourist trade. We saw it as a "sign" from Heaven and felt called to go minister to the poor people there, and then use that location as a steppingstone into central Bosnia.

Bosnia was a conglomerate of everything that made up Yugoslavia; a melting pot of religions, ethnicities, and nationalities. Under the reign of Marshal Tito, the mix was stirred up through cross-marriage and migration. But during the break-up of Yugoslavia, all the individual elements began to fight against each other, resulting in some of the bloodiest house-to-house fighting in Europe since World War II. We had seen many villages in both Bosnia and Croatia where the houses were all shot up, walls full of bullet holes or worse, and mine tape marking off houses or fields warning passersby that they had not yet been de-mined. But when we reached Sarajevo, it was the first time for us to see a major European city with half-demolished buildings, still occupied by the residents because they had nowhere else to go. The mother of one of our Bosnian volunteers was living in one such building, and we stayed overnight with her during our visit. She told us tales of how people would move from their apartments into the stairwells for more protection during the shelling, and how she would have to walk down the twelve flights of stairs when she needed water because there was no electricity for the elevator. Even if there was, she did not want to risk being in the elevator during a power cut. There were very few water sources, and she had to cross the open spaces and hide alongside the buildings to avoid being seen by a sniper while working her way to the water spigot. Getting back with the full jerry can was even more difficult, especially when climaxed with the climb back up the twelve flights of stairs. Looking down from her window, we could see the once grassy area below where children were meant to play now neatly cordoned off into little garden patches where the residents would grow their own vegetables. People survived in whatever way they could.

On our trips into central Bosnia, we found a large Croat enclave that enclosed twelve villages. One evening in December, we were driving in the Muslim area, and coming up through a valley when we came upon the next town. Lo and behold, it was Christmas! We had come into the Catholic Croat area, all lit up in celebration. The principal of the school invited us to stage our program for his children on our next visit, and he gave us a list of all the towns and villages in that enclave, along with the names of both the schools and their principals. This was where we were to move next, a quiet, Christian area that we could use as a base to then reach out to Sarajevo and the rest of Bosnia!

In Dubrovnik we had met a Croat man who had the key to the warehouse where they stored all of the aid and survival needs that had come down during the time of fighting. He took us there and said we were free to take as much as our van could hold to distribute to the people in Bosnia. Among the many things we took were a stack of large army-green tins of high protein survival biscuits weighing five pounds each. These biscuits were left over from World War II and had been hermetically sealed, so they would pretty much last forever. When we got them they were about fifty

Chapter 8: Post-War Yugoslavia

Young people volunteer show team

years old and in perfect condition.

We found out that there were also many refugee centers in Bosnia itself, and we began to visit them one by one. It was in one of those lonely centers outside Sarajevo that we met Mustafa and Lydia, who were refugees from Kosovo. No matter how much assistance we may have been in helping Mustafa and Lydia to rebuild their lives, we weren't really comfortable when they likened us to a true modern-day saint, Mother Teresa. At the same time, since Mother Teresa was of Albanian origin, it was understandable that this ethnic-Albanian couple would compare us to this lover and champion of the poor.

CHAPTER NINE

THE STORY OF MUSTAFA AND LYDIA

We met Mustafa, Lydia, and their children in the winter of 1998. Lydia was nearly due to deliver their fifth child, but was so thin and malnourished that she hardly looked pregnant. They had fled into Bosnia with their children through snow-capped mountains from Kosovo, which was having problems of its own in a struggle for independence from Serbia. They and other Albanian refugees from Kosovo were living in a freezing-cold, abandoned Coca-Cola factory just outside of Sarajevo.

The government of Bosnia, which was still recovering from their own devastating civil war, sympathized with their plight but could not offer them more than the damp, mold-covered walls to house them and a leaky tin roof over their heads. The same factory had housed Serbian refugees earlier, but they had been relocated because that area had been bombed with depleted uranium shells and had been declared unfit for human habitation. Many people in that area were reportedly contracting various cancers.

Update: 2013—We have read that more than 400 of those who were housed in that factory and in the surrounding area have since died of cancer.

These poor Kosovo refugees had to make rooms out of rough woolen blankets that had been provided by the UNHCR to form small cubicles for their families inside the cold factory walls with no heating in the midst of the Bosnian winter! Even those who fled the conflict zone continued to suffer the horrors of war! Everyone scrounged around the compound for bits of firewood, but no one was allowed outside the gates. Hundreds of grimy children ran around unsupervised. In short, the conditions were horrifying. We had gone to visit this makeshift refugee center to distribute humanitarian aid and offer the children our program and a chance to forget, if only for a short while, the trials they were enduring. A boy named Ed who spoke some English volunteered to help with translating. He was one of the few there who had taken any initiative and had started a makeshift school in the camp. We became good friends with Ed, and he was so thankful we had come. It was almost totally dark inside the factory, but it was much too cold to do our program outside. So we ended up performing by flashlight in a damp, gloomy corner of the building. We worked together to make the most of this difficult situation.

We did not have much time to get to know Mustafa and Lydia during that first brief encounter but were happily surprised to bump into them again a few weeks later when we were carrying out our humanitarian activities in a city in central Bosnia. They had been moved from the horrible conditions at the Coca-Cola factory to a "transit center" with slightly better facilities. It was at least moderately heated, but the air was so stale and humid that tuberculosis was rife. We began delivering aid to this center regularly, and each time we visited, we would prepare a special package for Mustafa, Lydia, and their children, especially for Lydia and their yet unborn baby.

The transit center was, however, only to be temporary, and they were soon moved again. The Bosnian government had finally finished building an "official" refugee camp where the ever-mounting number of Kosovar refugees could have better conditions. This camp, consisting of wooden barracks that housed up to six people per small room, was in an isolated area of central Bosnia, a three-hour drive from where we were living. By this time we felt like we had "adopted" this poor refugee family, and we would visit them as often as we could.

After some months, Lydia gave birth to a beautiful, healthy girl. It's a miracle that both survived, considering Lydia's frail condition. We continued bringing them whatever we could gather. During each visit they insisted on treating us like kings with what very little they had, usually a cup of the traditional Turkish coffee, biscuits, and sometimes juice. They had adopted us as much as we had adopted them. At one point, Lydia actually offered me the newborn baby, as she felt so thankful for our help when she was pregnant!

With the U.S. in the lead, NATO intervened in Kosovo in June of 1999. After those bombings, the refugees from Kosovo were told it was now safe for them to return to see what was left of their homes! On our last visit to their camp, we told Mustafa and Lydia that we were planning to begin humanitarian aid trips to Kosovo from our base in Bosnia and that we would try to see them there. Since there are no street addresses in rural Kosovo, all they could give us was their family name and the name of their town. Thankfully, in the villages, people pretty much know where everyone else lives.

Several months later, we were able to make our first trip to Kosovo. It was a sixteen hour drive over two mountain ranges on pock-marked roads that had been cratered by air-borne missiles and cluster bombs. The bridges had been bombed by NATO and, at one point, the traffic that included large trucks carrying timber for reconstruction from Bosnia would have to drive down through a field, cross a small ravine, and then back up the other side to get around a broken bridge. After a rain, it was a mess, and the cars would slip and slide up the muddy hill trying desperately to reach the road again. Some of the trucks would get stuck at the bottom and could not get out.

In spite of these obstacles, we drove on toward the capital Priština and found their village, the crumpled piece of paper bearing their family name in hand, hoping to find someone who knew them. The first person we met said he thought he knew where they lived and offered to show us the way. Even after working for two years in war-torn Bosnia, we were shocked by the condition of the area he led us to. It was utter devastation! We couldn't believe that anyone could be living in the shells of those houses that remained. After winding our way down some muddy back tracks, we pulled up to a high wall with peeling institutional green paint on it and banged on the metal door. A moment later, it was opened by a wide-eyed Mustafa smiling from ear to ear!

The reunion was a joyous one! We had come bearing as many gifts for the family as we could carry. Their house had been burned out, but they had managed to restore one room where they could all sleep and keep warm.

Mustafa and Lydia receiving aid

Chapter 9: The Story of Mustafa and Lydia

We continued our trips from Bosnia to Kosovo for the next nine months and then moved to Kosovo when the bulk of our work shifted there. We continued to see Mustafa, Lydia, and their children regularly. The baby that marked the beginning of our relationship was growing big and strong. They struggled to rebuild their house and lives, but we're thankful for the small part we could play in helping them recover from the tragedy that was the Kosovo conflict.

As part of the reintegration process, Spain had offered to take a number of Kosovar children on an exchange program for the summer. Concerned Spanish families were offering to take in children so they could learn Spanish and experience life in a developed western country, and the three oldest of Mustafa's and Lydia's children had been accepted into the program. At first, the United Nations Mission in Kosovo (UNMIK) had issued travel documents that Spain accepted, but the following year an official U.N passport was required. Mustafa and Lydia had put in the applications for the children's passports months earlier, but their papers were ignored and the children were in jeopardy of missing their bus that would drive them to Sarajevo and the flight to Madrid. We received a frantic phone call from Lydia asking if we could help the children get their passports. We had only two days, and our first reaction was "no way." Since the Lord had supplied so much for them through us, it seemed like they were now looking at us like we were agents of "God," able to do the impossible. Afterward, we felt bad about our original negative response and called back to tell them that we would see what we could do.

We had to start totally from scratch since we had no idea how the passports were being issued or what might be the cause of the delay. Apparently, the problem was in the local office that was burying their applications in the bottom of the pile, since Mustafa and Lydia were too poor to pay the "fee"—that is, a bribe—to the person overseeing that office. I knew we would not have any influence in that office, and we could not speak Albanian, so I asked around and made my way to a U.N. office to make enquiries. The very first person I met in that office looked at the name on my business card and was fascinated to see that my surname was the same as his former chief of human relations. He asked if I was perhaps his wife, and I responded that his former chief was my brother-in-law. This unbelievable "coincidence" was enough to win this man over, as he had greatly respected Andrew's brother, and he willingly offered to do all he could to make sure the three children would have their passports in time to catch the bus.

It took a few more phone calls and a bit of additional "pushing," but sure enough, with just hours left before the bus departed, we had the children's passports in hand! From the capital Priština, we made a mad-dash, two-hour drive to the town where the children were to load onto the bus headed for Sarajevo. We arrived just in time, handed them their passports, and wished them a very special summer holiday in Spain. During their three summers in Spain the children learned Spanish, won the hearts of their Spanish foster families (each child was hosted by a different family), and were blessed with the opportunity to experience things most Kosovar children can only dream of.

Update: 2013—After moving on from Kosovo we were no longer able to visit Mustafa and Lydia, and because they did not have internet, we lost contact. But I'm happy to say that we recently received a message to our website from their eldest son. He expressed his family's ongoing appreciation for all we had done for them to get back on their feet in those difficult years. He said the children were all growing up, and that his younger sister was now married with two young girls. And he asked if it would be possible to chat sometime on Skype videocam. We arranged a time and were able to see and talk to each one. The baby that Lydia had offered to give us was now fourteen years old! Since we were planning a trip to Kosovo in the coming months, we told them we would come to visit soon, and we did. It was quite an emotional reunion. We spent the afternoon with them, remembering old times, and taking photos. We had with us some aid and a box of Duplos (Legos for younger children) that had been donated by the Lego Foundation in Denmark on a recent trip to Scandinavia. We gave them to the two granddaughters. We are so thankful to have been able to be God's instruments of blessing on this precious family!

With Lydia and her daughters

Children playing with donated Duplos

CHAPTER TEN

OUR DROP IN THE OCEAN

"What we are doing is just a drop in the ocean. But if that drop was not in the ocean, I think the ocean would be less because of that missing drop."
— Mother Theresa

Helping others does not necessarily mean that you have much yourself. You just have a heart to help those in need, and are willing to share what you do have. So in addition to helping Mustafa, Lydia, and their family, we decided to concentrate our "drop" on a small Serbian enclave that was surrounded by Albanians who were intent on ethnically-cleansing the area. We had heard about this village from other international aid workers, as it had become well-known because of what had happened there. Having only 600 inhabitants, it had been the site of a horrible massacre where fourteen of the men had been killed in front of their wives and children as they were returning from working in the fields. Life in this isolated village is still dangerous to this day. Not long ago, a school teacher and her husband were killed while driving home. For fourteen years, the people have lived in constant fear, and what little protection NATO peacekeepers used to offer them has long since dried up. The poor villagers are unable to move to a safer area as there is no money.

The first time we visited this village it was the middle of the harsh Balkan winter. We had some banana boxes of donated clothes and things for children that we were hoping to distribute. We drove our van down the middle of the deserted, snow-covered, main road, stopped our van, opened the back doors, and then waited. Within minutes, curious people began coming out of their homes, eyes wide in amazement at the scene before them. They calmly began looking through the things that we had brought as the word had spread and more people began coming out to meet us. We did not have much on that initial visit, but what we did have seemed to multiply before our eyes as we simply leaned on God's supernatural supply.

After establishing a solid relationship with the residents of this tiny Serbian village and winning the trust of both the people and the village elder (who also happened to be the principal of the school), we decided to "adopt" the village, that is, to focus a significant amount of our time on helping them.

After several deliveries of aid, we felt a desire to do more for the young people of the village whose ages ranged from thirteen to seventeen years old. We started a regular program of character development where we would teach and train them in motivation, leadership, and various life-skills.

Our trainees receive their certificates

At the end of the training we wanted to do something special for them to commemorate their finishing of the course. We determined to take them on a day trip to a beautiful lake situated in a Serbian area near the Serbia border with Kosovo where they could enjoy a day free from the rigors of daily life, and enjoy playing in the water, boating, and other fun activities. The problem was, the lake was eighty kilometers (50 miles) away and required passing through hostile Albanian territory. We had to come up with some sort of solution for transporting the kids there!

Because the Danish NATO military contingent was overseeing the area where the lake was situated, we approached them with our idea and asked them for help. Their office for civilian/military cooperation (CIMIC) was open to helping us in any way that would enhance the lives of the minority Serbs, especially when it came to children. They agreed to supply two military transport trucks together with armored personnel carriers at the front and back that would guarantee protection for the convoy. This was important, as in those days, the radical element of the KLA (Kosovo Liberation Army) had an ongoing policy of ethnic cleansing of Serbs from Kosovo, and they would attack any sort of Serbian transport that was not protected. The Danes not only provided the needed transport but also supplied drinking water for the day and a military ration lunch kit chock full of all kinds of goodies, which the kids loved!

The military vehicles came down to the village in the early morning, and all thirty of us were assembled and waiting as they pulled up. It was quite a sight as we all climbed into the back of the green canvas-covered trucks and settled down on the not-so-comfortable wooden plank seats. Apprehensive parents handed cell phones to their children so they could keep in touch throughout the day and waved their goodbyes as the military column pulled out. For most of these children, it was the first time they had been outside their little village enclave in years, able to experience what "freedom of movement" was like—something that was supposedly guaranteed under the U.N charter.

Day of fun activities at the lake

After a day of great fun, the trucks and escort vehicles returned in the late afternoon to pick us all up and get us all safely back to the village before sunset. It was an amazing day and one of those unforgettable events for us all, but especially for the young people.

Since the lake excursion was so successful, we wanted to do even more for the young people the following year. We set out to plan a week-long seminar/holiday for our group on the Montenegrin coast. The distance this time was much greater—three-hundred-fifty kilometers (218 miles)—and while most of the driving would be in Montenegro, we had to catch the bus in the Kosovo Serbian town of Mitrovica. We would again require the help of the Danish KFOR for secure transportation from the children's village. Despite the successful trip to the lake the previous year, the parents again worried, this time even more, since it was a much longer trip, and their children would be gone for a week rather than just one day.

With Kosovo being landlocked and the Serbian village enclaves under constant threat, none of these children had ever been to the coast or even seen the ocean before. Most of them learned to swim during that week, and had the time of their lives as they took advantage of the evenings to walk to the nearby town to enjoy some semblance of night-life, discotheques, etc! Both food and funding to support this initiative were supplied via sponsorship raised from international organizations operating in Kosovo.

Summer on the Montenegro coast

Youth group special outing

School Construction and Football Shoes

Just opposite the main U.N. headquarters building in Priština, the capital of Kosovo, was the most popular English pub in the city called the "Kikri Bar." Many of the expatriates working in Kosovo, either for the U.N., a humanitarian organization, or a contractor, would flock there after working hours. For many of these people, it was difficult to live and work in a poor foreign country that had just come through a war. There were long electricity and water cuts, no garbage collection, etc. The pay was excellent, but the combination of poor conditions and being away from family and friends contributed greatly to their stress level. So sitting with colleagues for a few drinks after hours was just what the doctor ordered to relieve the stress, and was one of the few enjoyable pastimes available to them.

This bar was owned by a British man and his two partners. We had heard that they liked to occasionally help humanitarian organizations financially. I wanted to meet this man and have a chance to explain to him some of the activities we were involved in. The bartender told me when I could catch him, and we made plans to visit again the next day.

He was a rough, outspoken man, and not at all highly cultured like one perceives the British to be, but he had a heart for helping others. After hearing about our humanitarian activities, he said he would include us in the list of charities his business would support!

Each Tuesday evening, the pub hosted a charity-fundraising activity that most often was either a dart-throwing contest or a trivia quiz. Each patron that wanted to participate would pay an entry fee. The winner of the event would get half the pot, while the other half would go into the charity-fundraiser kitty that was distributed between the three chosen charities each quarter.

At one point, our new friend said he wanted to do something more just for us and suggested a charity auction. We had never heard of such a thing but thought it was a great idea! We set the date and went about town soliciting contributions that we could auction off. Several restaurants donated dinner for either two, four, or six people. One hotel donated a free night's stay. The pub itself threw in a case of wine. A shop donated a fruit basket, etc. When the day came, we had a pretty good list of items to offer at auction. The great thing was that these men (mostly) did not really care all that much for what they were bidding on. They just wanted to donate to our cause, and the auction was a fun way to do it. The fruit basket was actually auctioned off three separate times, as each time, the person who won the bid would put it back on the table so it could be auctioned off again!

The featured item to be offered at auction that evening was a free, round-trip ticket to any destination in Europe, donated by the regional manager of Austrian Airlines. That ticket itself went for more than 600 euros (almost $800 in 2013)! By the end of the evening the Lord had supplied more than 1,500 euros (almost $2,000 in 2013) for our charitable activities, and the pub owner asked us if that was "okay"! Meanwhile, one of his partners who owned a sushi restaurant offered to host a charity dinner as a fundraiser for our work a few weeks later.

During the auction, I was going around with one of our young volunteers to talk with some of the clientele in order to explain more about our humanitarian activities and why we were in Kosovo. One of the men we talked to who was in charge of a construction company that had lucrative contracts with NATO and the UN, said that he would like to help some needy cause, if we could suggest to him a worthy project. He invited us to his office, and during that meeting, we told him about the dilapidated, rundown primary school building in the Serbian village that we had "adopted" and were helping. He went to take a look at the situation on the ground and agreed to totally renovate the school, including a new roof and heavy wooden floor boards. They did an excellent job and were happy and thankful for the opportunity to put something back into the community. We found that many of the expats who are posted in these types of foreign missions feel very guilty about making such huge salaries while the poor local people suffer terribly as a result of the conflict they had just come through. We are happy that, in this case at least, one of them went out of his way to do a kind deed for others.

School reconstruction: Before

School reconstruction: After

After the completion of the reconstruction project, our friend the village elder/school principal, was understandably very thankful to us for all we had done for his village. Having seen how amazingly the Lord worked for us, he mustered up the faith to ask if we could help their teen boys' football team get shoes and footballs. We told him that we would see what we could do, but since it was the weekend, it might take a few days.

We were about to pull out of the village when, for no apparent reason, one of the doors of our van wouldn't close properly. It took us about fifteen minutes to figure out what was wrong. Just as we got it to close, we saw a Jeep with Finnish NATO peacekeepers approaching. We had a good working relationship with the Finns, and they had donated new classroom furniture for the renovated school, so I flagged them down to say hello. We talked about the village and its security, and eventually the need for football shoes came up. The peacekeepers suggested that we go to their base's CIMIC officer and explain the situation, and they phoned ahead to tell him we were coming.

At the base, we were amazed at how quickly things worked out. The officer in charge authorized a cash allotment for us to buy the shoes, and he personally donated two footballs. In the nearest Albanian town (shops in the Serbian enclaves would not have football shoes), we were shocked when we discovered how expensive the shoes were. Another obstacle was that the teenaged boys needed mostly larger sizes, but Albanians generally have smaller feet.

Finally, in Priština, we came across an official Adidas shop where they had all the sizes that were needed. The owner's heart was touched by the need and our story, so he lowered the total price so that it fit our budget! At the same time, he was happy to get rid of those larger sizes!

You can imagine the shocked amazement and huge smile on the face of the village elder when we came back after just one day with the footballs and shoes in hand. And the boys? They were overjoyed with their genuine Adidas shoes and knew they had been sent from Heaven straight to them!

It has been said that the happiest people in the world are those who have discovered the joy of giving, and we can attest to that! Whenever and whatever we give, blessings come back to us, often in the most unexpected ways!

CHAPTER ELEVEN

GANDHI OF THE BALKANS

Ibrahim Rugova was a prominent Kosovo Albanian political leader, scholar, and writer. Now called the "Father of the Nation" of Kosovo, he led the popular struggle for independence from Serbian rule and is considered by many to be the "Gandhi of the Balkans." In 1998, he was awarded the Sakharov Prize for Freedom of Thought, and has posthumously been declared a hero of Kosovo. He died in 2006 after losing his battle against cancer.

Known for the silk scarf around his neck that he was never seen in public without, swearing that he would not take it off until Kosovo had its independence, it was long suspected that President Rugova had cancer, although the truth was never made public until just before the very end. Committed to a non-violent path toward independence, Rugova was often at odds with the radical element of the KLA that was waging a guerilla war against the Serbian government.

A pacifist at heart, we had a feeling that he did not really want to be president. But the common people loved him, and NATO felt he would best serve the interests of the fledgling nation they had just helped to set up. And so it was that in his final days he struggled with both health and conscience. He lived in a simple house in the center of the capital Priština, and did not consider himself above others. We admired him greatly, and prayed for him often, as his position was a difficult one.

In 2004, we received an invitation to participate in a multi-ethnic children's day celebration that was going to be held at the U.N. Headquarters building. Since the conflict in Kosovo mainly centered around the differences between the majority Albanian population and the Serbian minority, the U.N. wanted to put on a "show" that Kosovo was a truly multi-ethnic success story, and they wanted to use the children as living proof.

They had selected children from several "multi-ethnic" village schools who would come to Priština and spend the day at the U.N. building attending various events. I say "multi-ethnic" in quotes because the children in these schools were not truly integrated in the classroom. They were kept segregated, with one group using the school in the morning, and the other in the afternoon. Still, it was progress in the right direction.

Knowing that we were specialists in multi-ethnic children's activities, high ranking U.N. officials asked if we would be willing to stage our program for the children as part of the day's activities. We had a really great time with the children, teaching them to sing our meaningful songs in both Albanian and Serbian languages, playing fun games, and presenting a message of reconciliation via mime theater!

Toward the end of our program, we were asked to wrap things up as the children were being invited to attend a press conference with the U.N. Administrator for Kosovo at the U.N. Press Office. Still dressed in our clown and animation outfits, we and the children sat patiently as we waited for the press conference to begin. Time dragged on, and the children were getting restless, so the organizers asked if we could do something with them to help pass the time. We jumped at the chance and took over the entire room, diplomats and press corps included, and led a rousing time of inspiration, singing and dancing around the conference room.

I dare say that room had never seen anything quite like that before! We had every video and hand-held camera focused on us as we marched the children up and down the aisles of chairs, singing our songs in both Albanian and Serbian. The following day we were of course featured on the local news!

Multi-ethnic day at the U.N.

After settling down and listening respectfully to the speeches, the children were led out to the U.N. parking lot where two large buses were waiting for them. Following a photo shoot, and not knowing what was happening next, we asked one of the officials where the children were going. He said, "First, they'll go to the amusement park, and then they will meet President Rugova at his residence." Our part of the program was now finished, but we could not pass up this golden opportunity to meet Dr. Rugova, so we climbed up on to the buses with the children. Thankfully no one seemed to question why the clowns would be going to meet the President.

We had many times walked right past Dr. Rugova's house. He kept very little security, and there was normally just one guard between the gate and the front door about ten meters away. On this occasion, the gate was open and welcoming us as we all filed out of the buses and were escorted into the room reserved for high-level visitors. Behind a large conference table and dais we could see on the wall two large framed photos: one of Dr. Rugova with Pope John Paul II, and the other with Mother Teresa. We knew this man was someone special!

With President Ibrahim Rugova

Sadly, he had just returned from The Hague, and looked very tired. There were the usual speeches and other boring formalities, but then his attention turned toward us, standing in the back, still in our clown outfits. He invited us to join him for a photo, and once the ice was broken and we had moved into the limelight, the Lord's Spirit took over. One of the songs we had taught the children that day was the Sunday school favorite "I've Got Joy, Joy, Joy, Joy, Down in My Heart." We knew that President Rugova had secretly converted to Catholicism, and so asked him to sing along with us in a rousing chorus of "I've Got the Joy" in his native Albanian. It was an incredible scene as the entire group joined in while the cameras continued to roll.

As we were leaving, he thanked us for the happy, joyful spirit that we had brought to his home, and we were able to encourage him in his difficult job! It was not long after that he passed away and went on to his Heavenly reward.

CHAPTER TWELVE

VICKY'S TRAVELOGUE

During our years in Croatia, Bosnia, and Kosovo, we always had several young volunteers with us to help us with our program and other humanitarian activities. These young people wanted to have the experience of living with people different from themselves and to invest their energies into helping those in need! They wanted to move out of their comfort zones and see what the real world was all about.

This is how sixteen-year-old Vicky described her first trip from Bosnia to Kosovo, just after the bombing:

"I suddenly woke up; my body jolted a few inches into the air as the van's wheels bounced through yet another pothole. My eyes jerked open, and I looked out the window.

It took a few seconds to register where we were. There were hills covered with trees. We were still driving through homeland Bosnia. Alas, we still had approximately fourteen more hours of driving until we would reach our destination: Priština, the capital of Kosovo.

Fourteen hours, that is, if we were able to enjoy smooth driving the whole way. No need to scare ourselves with the possibilities of roadside bandits, or corrupt border police who might take a liking to strange foreign women and keep them as part of their growing personal harem.

I closed my eyes again. Our journey had begun at four a.m., when we had pulled ourselves out of bed, piled into the pre-packed van, and headed off. (If only it actually was as easy as I just made it sound!) I shared the back of the van with Sharon, who had arrived from the far-off land of Australia only a few hours before our departure. Barely had she stepped off the bus when the details of her next trip were explained to her. After a meal, a shower, and three hours of sleep, she was whisked into the throes of our adventure.

Happily for us, the back of our van sported a comfortable bed, in which we two were quite content to finish off our night's sleep. Also along for the ride was our "almost human" dog, Joshua; being a dear "son" to Andrew and Anne, he had authoritatively claimed an unoccupied seat near the front. For our brave driver, Andrew, life was not quite so luxurious, but I comforted myself that along with this burden came the blessing of being the only male on the journey. He had an alert spouse at his side, so Sharon and I had nothing to fear as we lay down for a lengthy doze.

Now, at six a.m., we watched the radiant dawn illuminating the eastern sky, heralding the first day of our long-anticipated journey. Continuing toward the Montenegrin border, en route to our final destination, we soon found ourselves on the "main road." (This could more accurately be described as a rear-bruising, hair-raising, one lane mountain pass, hovering on the edge of formidable cliffs.) After nearly an hour of driving, we arrived at the border, a makeshift lean-to with a string stretched between two trees, and a rumpled border guard in pajamas. (Okay, slight exaggeration, but I assure you, the rest of the story is genuine!) The guard wearily looked us over, his face pressed against the window glass, peering inside.

"Passports!" he grumbled, then grunted as he studied our photographs.

Strolling toward the back, he grinned a toothless grin as he opened the door to discover two disheveled, semi-sleeping girls in the back. (A grin that sent those "female hostage" thoughts flooding through my mind all over again.) I heaved a sigh of relief as he closed the door, handed us our passports, and waved us over to the side. I later discovered that the necessary payment would have normally been a considerable sum, but after an explanation of the aid work we were doing, he let us through only paying about half what he originally was asking.

We spent the entire day traveling through beautiful Montenegro, with an occasional tank-up stop at a gas station, and for us juice guzzlers in the back to "tank out" at the resident squat toilets (or, more accurately, "dirt holes").

A number of times, we were halted by local police, who were eager to either demonstrate their superiority, or relieve us "rich foreigners" of a couple of German marks under the excuse of speeding or some other made-up offense. We found, however, that most of them were quite kind and maneuverable after a smile or two, and an explanation of our good works (sometimes with the help of a "gift" of some chocolate, or a pair of socks). Through it all, we only ended up paying money once, and that just ten German marks. This we cheerfully gave after a look at the recipient, and quickly categorized the money as "help to the poor."

Priština

I shall spare the innocent reader the microscopic details of our journey, lest this turn out to be a hardcover book, with captions on the back reading: "Two eyelids and a thumb down!" Suffice it to say: We arrived in Kosovo.

At that time, there was no border for entering the country. After fifteen hours of driving (or should I say napping), my rump skipped a beat, and I invented a soothing song titled, "We're Almost There."

All around I could see the most obvious effects of the recent war: broken buildings, rubble, and roofless houses. A key bridge had been bombed out, so we waited for an hour as the line of vehicles ahead of us slowly inched their way into a rocky ditch, then up the slope leading back to the road (which, though far from smooth itself, felt like silk compared to the previous one). I watched as the truck ahead of us inched on, then tipped precariously to the side as its right wheels sank into a pothole. To my relief, it managed to straighten out again, but that relief was short-lived when I realized that it was now our turn.

I held my breath, praying, reciting the 23rd Psalm, and solemnly promising never to be bad again. The praying must have worked because we survived and finally ended up back on the road, and a little closer to our earthly destination.

The rocking, bouncing van rolled into sweet stillness, parked amid a row of vehicles on the side of the street. I opened the curtains and peered out the window. The street, closed for pedestrians only, was fairly crowded with stylishly dressed, laughing and talking groups of young people strolling about. A feeling of freedom and joy permeated the air. Somehow it just did not match my expectations of how I thought this city, so freshly emerged from a war, would look.

Across the street was a medium-sized building, the large letters at the top heralding "The Grand Hotel." It was the spot where we had prearranged to meet up with the team who had already settled in the area. We heard laughing voices as they approached our van and greeted us with warm smiles and hugs. They instantly struck up a conversation with Andrew and Anne, explaining that the rest of their team was at a local concert and would be here shortly.

We crossed the road and entered the hotel, in hopes of using the bathroom. Five stars proudly stood at the top of the building, but we soon discovered the regular "squat toilets" we had encountered at Podunk gas stations along the way. Despite its supposed higher standard, this bathroom would be lucky to get one star. I eagerly searched for a mirror, dreading the sight I knew I would find, but was spared from the view, since there was nothing on the raw cement walls. No water came from the taps either ... oh well!

In the lobby, the setting was presentable, however, and as we later learned, the hotel had only recently been returned to its original use as a guest hotel, having first been used to house refugees, and then for members of visiting NGO's (non-governmental organizations).

Back at the van, a small group emerged from the throngs of weekend youngsters, and we greeted the rest of the team: David, Michael, and Esther. They were going to help us with music, acting, and translating in our upcoming programs in different institutions.

The men drove the vehicles to the nearby student dorm where the administrator had so kindly agreed to let us stay for free. The rooms were simple but nice, each with its own bathroom, and yes, a MIRROR (groan…). We discovered that for the last few days, the whole city had been plunged into times of frequent waterlessness (okay, so it's not in the dictionary), and we had to grow accustomed to the timings of its sudden and wonderful appearance.

Here are some facts and figures about the culture and surroundings:
- It seems most of the drivers here are crazy; maybe it's a manifestation of the traumatic experiences they've been through or merely a desire to put their recently attained freedom to use. Most likely, the reason for this uncouth driving behavior is the disorganization of things in general, including the lack of traffic police or working stoplights. (The cars don't need license plates and drivers don't need licenses.)
- One must always be aware of the many buried land mines that are just waiting to be stepped on. Numerous innocent people have fallen victim to these terrible remnants of the war. Because of this, we only walked on the streets, not stepping off the sidewalk unless the ground had obviously been stepped on recently.
- Another outstanding factor is how many foreigners are roaming the streets, either standing guard or in matching army jeeps with KFOR (Kosovo Force) heralded on the side.

The general feeling of the people toward the many British troops is good. You often see people waving at a passing jeep filled with soldiers. International police forces are stationed here as well; you sometimes will see a sweaty foreign policeman in the throes of guiding traffic and trying to comprehend the Balkan ways. Besides these peacekeepers, there are swarms of aid agencies with their names plastered on their minivans and trucks. There is no doubt we looked somewhat out of place, driving around in our rusty, rainbow-curtained van.

"Don't talk so loud …"

Most people here speak English quite well and will go to any lengths to be hospitable. The majority dress quite fashionably, not too far to either extreme as far as being conservative or revealing.

According to the locals, there were strict eight o'clock curfews before the war. So this new era of freedom comes as a great joy to the people, and evenings are a time for celebration. Bars are open. Discos, however, still remain closed.

Resentment against Serbs runs extremely high. Speaking the Serbo-Croatian language is forbidden and greatly resented (which we had to adjust to, as our programs were previously in Serbo-Croatian). According to one of our friends, they are trying to enact a law that would fine anyone overheard using the Serbian language! Signs bearing Serbian words were spray-painted over, and the Albanian translation scrawled over it. Previous Serb residents have been kicked out, and some of their houses burned. KFOR soldiers guard remaining Serbian Orthodox churches twenty-four hours a day against possible vandalism, or worse. Some were blown up!

I was talking with a young guy, and, after I told him I live in Bosnia, he asked if I spoke any Serbo-Croatian. I proceeded to rattle off a few words. Alarm quickly came to his face as he darted his head about, his eyes scanning in each direction. He spoke in a hushed whisper: "Don't talk so loud! If anyone hears you …"

Along with dislike of Serbs comes a high nationalistic pride.

On one occasion, a friend invited us to a student party. Once inside, we were instantly overwhelmed by the traditional Albanian music booming in the crowded room. The elegant and fashionably dressed teenagers were in a massive line doing a simple national Albanian dance routine. Right leg, left leg, cross, step, right leg, left leg ... and on and on.

It looked like something I would have expected from a group of white-haired, traditionally costumed older men and women on a national holiday. Yet, here was a crowd of teens and young adults, dressed more appropriately for a stylish disco, dancing the national dance their ancestors probably wouldn't have done when they were their age. According to our Albanian friends, none of them would have done it either a year ago, but, now, they did the same step repeatedly, song after song, with a pride and bounce to their steps. I eyed them curiously for a few songs, until I was invited to join in. The step looked simple enough, so I clasped hands and began the dance. Soon the room grew so crowded with aspiring dancers that my toes began to ache considerably from being trampled on, not to mention my arm becoming semi-dislocated from being pulled every which way. The dude on my left was so distracted by beautiful Esther, who held his other hand, that he began to lose the beat completely, and I found myself smashing into him every time I followed the synchronized line on the right of me doing the step, cross, step. Oh, well, at least I can say I did it!

Clowning up

We came here to help the children recover from the trauma of war, so we put together a program of positive songs, theater, magic tricks, games, and meaningful skits. The performances were enjoyed almost as much by us as by the audience, perhaps because of the adventure that comes with knowing one is about to come this close to a real-life saw pretending to cut you apart, or a pan flying towards your head as a professed anesthetic. Such was the case in our "Doctor Skit," in which a patient is relieved of the clutter ailing his heart, and in its place is put love, joy, kindness, sharing, etc. One noteworthy downside to having to "clown-up" each day is the "spaghetti face" that comes from a painted smile stubbornly insisting on leaving a reminder of itself there, which leaves us hoping it will remain as determinedly in the hearts of the audience.

Vicky (2nd from left) ready to perform

There was no electricity, so we had to have our own generator with us. We had to have an extra long cord as the generator was so noisy we could not be heard otherwise! It was rather comical!

Time would fail us to tell of all the other adventures we encountered. My mind would also suffer a major breakdown trying to compile them all in a decent, readable way. So you will just have to believe me when I say we had a truly marvelous, crazy, interesting time, and we will move on in an attempt to make this short ... (right!).

Gjakova

The day we left Priština, the streets were crowded with flag-waving citizens, triumphant and nationalistic. It was the NATO deadline for the disarming of the KLA. The troops paraded the streets, along with throngs of people honking horns, singing, and shouting. Amid all this stood our van, inching its way through the traffic, biding our time with an occasional wave out the back window at some heroic soldier.

With no traffic lights working and the streets thronged with people, we finally made our way to the edge of the city, and after a few hours of driving, we reached Gjakova, a medium-sized city on the west of Kosovo. Other members of our team had gone on ahead to seek out room and board for us: a big, empty farmhouse! Yeah, that's right! Cows, pigs, flies, and all! The dear friend who gave us free use of the place informed us that it had not been occupied for ten years. It smelled like it, too!

The city of Gjakova was about 70% destroyed, with a vast number of buildings burned. There was so much destruction! We did our performance many times and passed out the remainder of our humanitarian aid to these needy, desperate, and extremely thankful people. Many have nothing left and live in tents or shacks.

One evening, a friend led us down a dark alleyway in the center of town to a huge heap of rubble. This, he explained, had once been the town center and main hangout. Now, it was a ghost town. His eyes grew moist as he recounted all the things he had done in this very spot. The majority of his memories had happened where we now stood—on that heap of broken glass and concrete. As an observer, it's sometimes easy to overlook the great personal effect that this war and the NATO bombing has had on countless people who have lost family, friends, homes, and so much more.

We spent our last night in a former hotel, now a refugee center operated by sweet young Czech volunteers. They generously let us use their clean, simple rooms. We slept on mattresses on the floor, since there was no furniture. But we were thankful and happy that this, our first visit to Kosovo, had been safe and exciting!

CHAPTER THIRTEEN

MIND-BLOWING MIRACLES AND UNUSUAL CIRCUMSTANCES PART 1

Having spent fifteen years living full-time in the various republics of the former Yugoslavia, there are so many amazing stories still to be told. Sometimes I look back at old photos or read my old journal entries and think, "Did that really happen? Did we really do that?"

Before moving on, I want to share a few more of these real-life events with you in this chapter and the next.

Keys to the Containers

We often had cooperation with different NATO contingents in Kosovo. One of our closest relationships was with the British. They knew about our charitable activities in the field, and one day, one of our officer friends told us that they had just received four containers full of donated items that had been collected by school children and churches in Scotland. They had no idea who needed it most and no time to give it out. They drove us to where the containers were parked and gave us the keys! No matter how much we gave out, there was somehow always more. We were always very generous with our neighbors and the people in our area, but we spent most of our time in small villages and mountain enclaves.

Do You Need Any Food?

We would often be invited to eat on different military bases. After one lunch the chief cook came up to us and asked if we ourselves needed any food. Since we live by faith and trust the Lord for all our needs, our answer was of course "Yes!" He said "Give me a few minutes", and when he returned, he had put together a small pallet of all sorts of foodstuffs from the military pantry. Then he said "I didn't know exactly what you needed, but if you bring me a list next week, I'll give you everything you ask for." His giving continued throughout his tour of duty in Kosovo, and when he left, he introduced us to his replacement who continued with the weekly pick-ups. Much of the food in Kosovo is unsafe for many reasons, particularly because of the use of depleted uranium shells during the bombing. That's why the military always flies in its own food. So we were fed like kings, or to be more precise, like ambassadors – God's ambassadors!

Burned Houses, Warm Jackets

In 2003 there was anti-Serbian rioting amongst the Albanian population as part of the ongoing ethnic cleansing, and many poor Serbs had their houses burned in the process and were now living in very difficult conditions in adapted shipping containers donated by the U.N. An international friend of ours, a businessman who ran shops on the military bases, wanted to help these displaced people and donated hundreds of nice, good quality winter jackets, warm woolen and thermal blankets, sweaters and hats for kids, etc. A base that was upgrading their china gave us all their used plates, and another gave us stacks of cleaning materials when they were packing up to leave Kosovo. There were no jobs and constant insecurity. People had lost more than their homes; they had lost their land from which they survived. No more gardens, no more animals, no more life. We knew the bits of aid that we could bring were only a drop in the ocean to the overall need, but the encouragement felt by the people when they saw someone cared, was worth so much to them during their darkest hour. The need was always great.

Emotional women receive new coats

Distribution of blankets and coats

I've Taken a Collection for You

One of our military friends approached us one day and said, "The chaplain on our base has been taking up a collection each Sunday during his stay in Kosovo. He's been here for 6 months, but now he's leaving, and he didn't know what to do with the money, so he asked me if I knew of a good charity to whom he could give the money he'd raised. I recommended you guys." He handed us an envelope, and we thanked him politely without looking inside. Once home we opened it and were amazed to see 2,000 euros (over $2,600 U.S.)!

Snowsuits from Sweden for Christmas

During our visits to the multi-ethnic town of Janjevo in central Kosovo, we experienced some of the most deplorable housing we had seen. Roma gypsies were living in deserted, broken down houses without windows or heating. As we drove by, their cute kids would hang in doorways or sit on the front steps, and it really touched our hearts! It was winter, and the rags these kids were wearing were filthy as well.

We had some friends in Sweden who would often receive end-of-the-line donated items from department and chain stores, so we wrote them to ask if they might have any warm winter clothes that we could give to these poor kids! They told us they had very nice snowsuits. All we needed was the transport to get them down to Kosovo. We went to the appropriate NATO office to ask them about shipping some boxes of snowsuits for us on their next supply flight. To get this okayed would require someone who would be willing to go around standard military policy that forbade the shipping of civilian goods on a military flight.

The day before Christmas, we got a call from the base letting us know that the boxes had been shipped and would arrive for Christmas. In addition to the snowsuits, there were hygienic items and toys. We filled up our van and headed off to Janjevo like Santa Claus to give out our presents! That was an unforgettable Christmas, as we were blessed with the opportunity to put these Swedish quality snowsuits onto shivering Roma gypsy children. Months later, with the snow melting and spring around the corner, we could drive up to Janjevo and see the children playing, still wearing the same snowsuits, now looking like they had come through WW2. It seemed they had not taken them off since we put them on them. But at least for that winter, they had been healthy and warm!

Why Won't the Car Start?

We were driving from our home to the center of Priština to do some business, but we needed to first stop in at the Swedish NATO military base. The car had started just fine when leaving home, but after about twenty minutes on the base, when we turned the key to start the car, it was dead: no clicking, no turning over of the engine, nothing. Just silence. Andrew got frustrated since the car had been starting fine and there was no "physical" reason he could think of why it would not start now. I suggested he look under the hood, that maybe a wire had come loose. But he resisted, saying, "How could a wire come loose while the car was sitting still?" Finally, when he had no other recourse, he backed down and popped the hood. The first thing he noticed was the wire to the coil had fallen off. Huh! He put it back on, and, vroom, the car started like a charm. Proceeding on our way to the city, we reached the outskirts and saw that the traffic was being diverted. We followed the detour and tried to make our way to the parking lot where we always park. It was inaccessible, as there had been a bomb blast in that very spot only 15 minutes before. Now, how did that wire fall off?

Teacher Training Seminar

We had a very good working relationship with two ladies from Save the Children. They had projects of repairing and renovating kindergartens, and when each one was ready to open, there would generally be a dedication ceremony with school administration and local government officials in attendance. Our part was to stage our children's program with fun action songs in the local language, games, and meaningful theater. One day, while visiting the ladies' office, I overheard someone talking on the phone about how they had an allocation of money designated toward multi-ethnic teacher training, and if the money was not spent by the end of the month, they would lose it. They had the Albanian and Serbian teachers and school directors chosen to attend the training, but they had no one to present the seminar. I told him we could do that, since we were specialists in early education, and they were very thankful. It was a three-day affair that would take place in a posh lake-front hotel in the touristic town of Ohrid in Macedonian, and it would pay 400 euros each for three presenters, plus generous pay for our local Serbian translator who was one of our trainees. The venue needed to be in a neutral location, to avoid any ethnic conflict between the attendees. At first, the two groups flocked together, sitting as a block in the sessions and at meals. On the second day, as people came in, we asked them to draw a number from the hat and to please sit in the designated place, therefore mixing the groups and breaking down the walls of division. Soon they were voluntarily sitting with others from the opposite side, chatting together at meals, and learning that we are all the same at heart.

Multi-ethnic teacher training seminar

Multi-ethnic Mountain Camps

In Kosovo, multi-ethnic activities were high on the agenda of the U.N. We were able to participate in two camps for children, one summer and one winter, in the beautiful Shar Mountains that form the border between Kosovo and Macedonia. Our responsibility was to supply fun training for the children in the form of meaningful theater, interactive songs in both Serbian and Albanian languages, games, and other activities. The goal was to encourage the children to participate wholeheartedly in the activity without regard to the other children's ethnic group. There were about two hundred children each week, and they were divided into smaller groups of about thirty. They wore different colored caps in order to identify which group they belonged to. We brought several of our trainees to help us with both interpreting and animation.

Multi-ethnic mountain camp

The trainees did great. In fact, the U.N. overseers voted them best teachers at the camp in spite of their young ages of sixteen to seventeen. This was much to the chagrin of the secular teachers who had come for the purpose of helping train the children. They ended up spending most of their time sitting in the lounge smoking cigarettes and drinking coffee at the expense of the organizers. There was a bit of jealousy as a result, and our young people were even challenged as to whether they should have been allowed to attend due to the fact that they had not yet turned eighteen. But their zeal and enthusiasm more than made up for their lack of years.

The Ravioli Circus

Ravioli Circus in Kosovo

One spring we received a message via our website from two international drama students who were studying in New York City, a boy from France and a girl from Hungary. They wanted to come during the summer to do charity clown shows in Kosovo villages. They were looking for someone to take the lead in booking the shows and organizing their schedule, and we were happy to help them. They called their show the "Ravioli Circus," which incorporated a clown routine with magic, as our team took care of the interactive songs and games. They had learned quite a few of their lines in both Serbian and Albanian and also used Italian during their show. The girl played "Mama Ravioli" with a big pillowed tummy, and the boy played her son, Giovanni, who towered over his small-statured mama. They did quite well and kept the children in stitches, regardless of whether we were playing in the Serbian or Albanian area, in the center of Belgrade or a small ethnic village. We did not have any sponsorship to cover the costs, but the Lord supplied everything we needed, be it food, lodging, or transportation. They returned to their studies inspired and rejuvenated after spending nearly three weeks in a war-torn region, helping to change young people's lives.

Set Free in Prison

The sixteen-year-old son of our neighbors had gotten himself into trouble by getting drunk and breaking the window of a local shop. The U.N. police picked him up and put him in jail, and his mother came to us to ask if we could help him. We heard that even though he was a Serb, he was going to be soon transferred to the large Albanian prison that was for more hardened adult criminals. Surely, this would not be good for him. Because the U.N. was overseeing the reconstruction of Kosovo, the current director of the prison system was a Finnish lady who had been appointed by the U.N. She had become our friend previously when we were doing prison programs in various locations. I went to meet her and explained to her that we were friends of the family of this young man who had made a stupid mistake and was now in jeopardy of having his life ruined by being sent to that prison. She was sympathetic and said she would look into it and, hopefully, get him released soon. I had a thin, inspirational book with me that talked about how obstacles in life are to be overcome like an athlete jumping over hurdles. I asked if I could get it to the boy. She said yes, as long as it was not political or pornographic and she could deliver it to him herself. It took a few days before his release could be arranged. We heard from him afterward that he had read that thin book several times, and that it had changed his life. He was sharing a cell with other boys, and they each began begging the book off of him so they could read it for themselves. This boy's life did change, and he is now grown, married, and has two beautiful children and a good job in Belgrade.

Where's Our Car?

We were on our way to visit the American NATO base for eastern Kosovo in our aging Volkswagen Golf. Because of the location of the base, it was difficult to find a place to park. We ended up asking a gas station attendant across from the main entrance to the base if we could park in his station, to which he agreed. After spending an hour or so on the base, we came out to find our car was gone. At first, we figured it had been stolen, since there is a lot of crime in these regions that have been in a war fever, but we decided to ask the guard detail at the entrance to the base if they had seen anything. They said, "Oh, was that your car? We were afraid it was a car-bomb, and we called the police to come and take it away. You should have told us. You're lucky we didn't blow it up right where it sat!" We had to make our way to the police impound lot on foot. Thankfully, a U.N. police vehicle happened by, and we flagged him over. He was very friendly, and after explaining our situation, he found out where the police lot was located and agreed to drive us. He talked to the attendant, and they let our car out without charge. Whew!

Flying on 9/11

Andrew sometimes flies to the U.S. to visit friends, relatives, and long-time supporters. In 2001, he had a flight booked for September 11, a date that has gone down in infamy. But this is his story, so I will let him tell it.

From Andrew:

I was flying out of Budapest, Hungary, on Delta, while Angelina, because a friend had given her some free air miles, was flying on Air France via Paris. It was a normal transatlantic flight up until the time that the pilot came over the intercom to tell the passengers that he was receiving some news on the radio that could affect us, and that as soon as he had more information, he would pass it on. After about fifteen minutes, he came back on to say there had been some sort of "incident" in the States, and that the air space had been closed to all traffic. Our plane was being diverted to Newfoundland, Canada. We did not know at that time what the "incident" was, but we knew it must have been serious to divert international flights. At the same time, Angelina's Air France flight was being told to turn around in the mid-Atlantic and return to Paris. Non-U.S. carriers were not getting clearance to land in Canada.

We had been told to land at the international airport in Gander, Newfoundland. I peered out the window, and as we circled, I marveled at two things: the beautifully odd terrain and over forty planes that completely covered every available piece of tarmac. Once on the ground, we were not allowed to disembark because of security concerns. The planes had to be unloaded one at a time, and the passengers were then taken through stricter-than-normal screening upon entry. We watched as a set of four yellow school buses went from plane to plane collecting the passengers and wondered how long it would take before they would come for us. Since our plane was one of the last to land, we ended up waiting for a total of twenty-two hours before they finally got to us. By that time, there was no more food on the plane. Even the peanuts and pretzel packets were gone. The in-flight movie had been "Shrek." We watched it three times, but we still weren't "there" yet! I was thankful for an on-board satellite phone and managed to call Anna. She was obviously relieved. On the other side of security, the Red Cross was waiting for us with tables full of sandwiches and drinks, and free international phone calls were available so passengers could call anxious loved ones. I called Anna again, as well as my Mom, to let them know I was okay.

Housing the refugees was a huge problem for the Canadian authorities, as 13,000 people had literally dropped into Gander that day, a town of only 7,000 inhabitants. The passengers from my plane were bused to a small town forty miles away. They had prepared army cots in the high school gymnasium for us. As our buses pulled up outside the community center, I felt like I was on a team that had just won the Super Bowl! It seemed everyone in town had come out to welcome us, and they formed a column that we walked down through to their cheers and pats on the back. The way the Canadians responded to this emergency was nothing short of fantastic!

The high school had opened its computer lab so we could use the Internet, and while I was writing an e-mail, a lady poked her head in and said she could take four people to her house to take a shower. I jumped at the chance! It was there that I saw on TV for the first time the visual images of the Twin Towers collapsing. We chatted over a snack, and when the subject of the sleeping arrangements came up, I mentioned that I had a bad back and could not sleep well on an army cot. A few minutes later, the lady came up to me to say that she had called her friend who ran the Boys and Girls Club in town and that she had offered to let me sleep in the bed that was in her office! And, oh, there was a computer there with Internet as well! I knew Jesus loved me!

During our five days in Newfoundland, meals were provided by the local community. Residents brought in pot luck dishes and set them out on the tables provided. We all got to know each other fairly well, and pretty much everyone on my flight heard sooner or later that I was on a mission from Kosovo. The local churches arranged an inter-denominational memorial service for Sunday. The pastors invited me to offer a prayer for the families of those who had lost loved ones that day. It was a very moving experience.

All that time, we really did not know how long we would stay in Canada. Then, finally, we heard at dinner that the yellow buses would be coming for us the next morning. Someone suggested that we have a five-year reunion in that tiny town for anyone who would be able to return to remember those amazing days that we had spent together.

What a cheer went up when our flight finally touched down safely on U.S. soil. A tragic event had transformed 250 strangers on a plane into good friends who were hugging and crying as they disembarked. And Angelina? The Lord took care of her as well. She stayed five days with friends in Paris while she waited for her ongoing flight and arrived the day after I did.

In the midst of such a tragic event, it can be easy to become discouraged or to dwell on the negative. Thankfully, we were not on one of those four planes! Looking back on it all now, I realize that God gave us a blessing during some very tragic and difficult circumstances. It is an experience I will never forget. I believe one thing 9/11 showed was that during the most trying circumstances, the true, sacrificial nature of man will come to the fore.

Note from the Publisher: While this book was being prepared for publishing, Andrew was again flying to the United States during the time of September 11. On this occasion, his connecting flight from Newark airport was cancelled due to severe weather, but he made good use of the overnight delay, comforting fellow passengers by telling his story of flying on 9/11. We thank God for His protection and commend Andrew on his faith and courage.

CHAPTER FOURTEEN

MIND-BLOWING MIRACLES AND UNUSUAL CIRCUMSTANCES PART 2

Bullet-Riddled School in Sarajevo

In Bosnia, one of our primary activities was staging programs for children. At that particular time, we had the privilege of having a professional clown cum magician visiting us, and we learned so much from him. The very first school we went to was one of the oldest, most famous schools right in the center of Sarajevo. Upon arrival, we were moved by the sight of the outer facade of the building, which was riddled with bullets.

As we climbed the stairs on the way up to the room where we would be performing, we saw many bullet holes in the stairwell

In front of Sarajevo school

windows. Inside, the school was also in a serious state of deterioration, but the children were so happy and thankful for our coming and were waiting for our program with excited anticipation. The teachers and the director were so appreciative. The school had just recently re-opened, and after years of no school and the entire population of Sarajevo living in terror, our presence was a breath of fresh air and brought much hope for the future.

Living on the Front Line

Saša was a Bosnian Serb who grew up in Sarajevo. His parents lived in a fourteen-floor apartment building that found itself right on the front line between the Serbian and Muslim sectors of the city so that the end of the building facing the firing line was totally shelled and crumbling. They lived the entire four years of the war in this embattled building, and when there was shelling, people would flee from their apartments into the inner stairwell for protection.

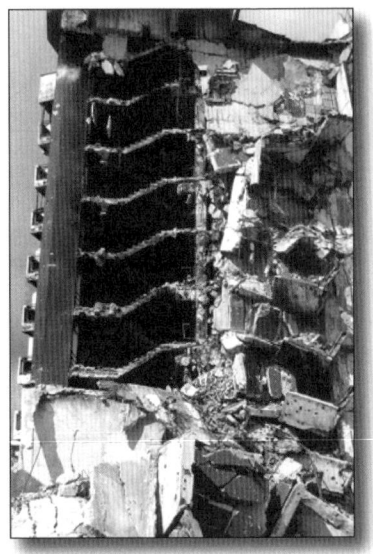

Destroyed end of apartment building

Neighbors comforted each other and often held prayer vigils or Bible reading during these traumatic times.

Saša's father would go out, sometimes for days, and walk up into the surrounding hills where there was a serious risk of land mines to pick naturally-growing herbs, something Serbs are very knowledgeable about. He did not want the horrors of war to keep him from his connection with nature.

They were kind, friendly people, and received us gladly when we visited them. Andrew was about to get into the elevator when Saša warned him not to. Even though the lift was working, the power cuts were very unpredictable. It was better to walk up the fourteen floors.

Saša himself was based in a village outside of Sarajevo. We had several barbecues where Saša's parents were able to come. They had such a good time tending to the grill and dancing to traditional Serbian music, enjoying their time outside the city. Sadly, both parents are now suffering from cancer, a fate that plagues many of those who survived the war but were exposed to carcinogenic depleted uranium and other contaminants during those years.

Watch Out for that Tank!

Driving in Bosnia in the early days was definitely not for the faint of heart. The roads were already badly deteriorated and had been further chewed up by heavy-treaded vehicle usage (tanks, armored personnel carriers, etc.). Alongside the normal- to large-sized potholes were larger craters that had been caused by air-to-ground missile fire from NATO. Some of these craters were so large you could actually drive down into them and then out the other side. Avoiding these holes was like driving an asphalt slalom course. Zig left, zag right, let that one go between the wheels.

Angelina by NATO tanks

CLONK! Oops, hit one. It's a miracle the undercarriage of our vehicle never collapsed on us in exhaustion. Then there were the tanks that took up two-thirds of the road. When one of those babies was approaching, you bailed out quick. Tanks were known to sometimes just drive right over a car if it was blocked in.

Once day in Kosovo we were totally stopped in a line of cars waiting at a checkpoint when a NATO treaded vehicle was passing the line. He must not have had that much room to pass, as he scratched against the side of our van. It was not a lot of damage, but we went to the base to complain. We were told that NATO was not responsible for any damage caused by their heavy vehicles. Oh, well.

Mine Fields and Freezing Temperatures

When we first arrived in the village in central Bosnia where we would live for two years, we drove down the main road that passed through the village and were shocked by the amount of "mine tape" that was put up. This area had been hotly contested during the war, and many fields and yards had been laid with land mines, and houses had been booby-trapped. The international relief effort had come and was in the process of de-mining the area, but it was a slow process. We stopped our vehicle and got out to take a few photos, and one of the de-miners, a young man from Canada, asked if we would like him to take the photo. He invited us into his field trailer and we made friends with him and told him that we were looking for a house to rent. He had a local girlfriend, and she told him about a house nearby. On our next visit, we went to see the house. We rented it, even though village houses in Bosnia had no central heating and only one pot-bellied stove in the living room. We had a team of six young people volunteering with us at the time, and they had their rooms upstairs, but those rooms would get so cold that they would come downstairs and sleep around the hot stove, taking turns waking up to add more pieces of wood. Our bedroom, which was downstairs, also had no heating. One morning, we woke up and found a glass of water we had left by the bedside was frozen solid!

Chapter 14: Miracles and Unusual Circumstances, Part 2   121

Frigid Bosnian winter wood pile

NATO's Dumping Ground

The place where we lived during our four years in Kosovo was only about a kilometer from a drop point where NATO bombers dumped their unused munitions before returning to their base in northeast Italy. When we were living in central Bosnia, at night we would see the lights of the squadron of bombers eerily flying directly over our heads on their way to deliver their payload over Kosovo. If for any reason they did not or could not drop all their bombs, they were not allowed to return to base with them because of the risk of landing with armed missiles, etc. So their instructions were to drop them in a designated spot. That spot was in an uninhabited hill just outside Priština, but very close to the Serbian village where our team was based. We were told harrowing stories by the locals that during the actual bombing and the bomb drops, people would run for cover, but none of the humble village houses had basements. There was not really any good place to hide. The only house that had a basement, because it was a larger 2-story house, was the one we had rented. The owner, who was telling us the story, showed us the basement. It was very small and damp, only about 160 square feet in size and more like a potato cellar than a real basement. He told us that up to twenty people would run to his house during a bombing raid, and they all huddled together down there until the bombing stopped.

Even before we left Kosovo, I started having problems with my stomach and digestive system. As a result, I had less and less energy, and sometimes would nearly faint. I was very sick most of the time, hardly able to stand. I spent the majority of my time lying down. I could no longer tolerate foods that I had never had a problem eating before, like meat, bread, or anything made with yeast. All I could handle was boiled fish with boiled potatoes and carrots. That is all I ate for about three years. I realize now that I had become sick from exposure to the contaminated ground and dust from this NATO dumping ground. Now, I often read of reports from that area that almost immediately after the bombing, the cancer rate began to go up dramatically, and that now, fourteen years later, an extremely high number of residents are dying of cancers linked to that contaminated area. Thank God that through prayer and a total change in my lifestyle, I recovered fully. For the past few years, I have been able to eat normally, work, walk, and even exercise again!

Too Cold to Come

During that same winter, we had been doing many Christmas programs and giving out aid in our area. In one particular town near us, we had been invited to do our show in their large community center. All the plans were made, and notices were posted around the town (stapling posters onto telephone poles and bus stops). Even the TV station was going to come to cover it. The day of the show, however, it was so cold that no one showed up except the TV crew. We delayed for half an hour, shivering in the unheated auditorium and hoping some people would show up. Then we decided to go ahead and perform for the TV. God doesn't judge us by our successes or failures. He only looks at our faithfulness to share His love with others.

Downpour Border Crossing

The Catholic charity "Caritas" had asked us to come to their storage in Croatia to pick up aid to carry back to Serbia. Their warehouse contained a mountain of banana boxes and large black plastic bags full of donated brand new clothing. Some years had gone by since the end of the conflict in Croatia, so there was not as much need locally, and they wanted to help the poor people in Kosovo who were in a desperate situation. Our van was packed full, side to side and top to bottom, and we were uncertain how the border customs would handle it. The folks at the storage facility had given us a print-out on their letterhead stating that the contents of the boxes was humanitarian aid, but that was not in any way an official document that would satisfy a picky customs inspector. As we were approaching the border area, we prayed for protection and for a miracle to get this aid to those most in need. The sky had been threatening rain for a while, but nothing had fallen. About a minute before we reached the border, a torrential downpour started to fall. We pulled up to the small border crossing where the officials sit in an adapted shipping container office and stopped at the lowered gate. One of the men looked out the barely open door, not wanting to venture out into the heavily falling rain, and just waved us through without even bothering to look at our passports, much less inspect the contents of our vehicle.

Anything on TV?

Our friends at the British base told us that one of their smaller outposts on the other side of the city had recovered a stolen TV, and they asked us if we knew a needy school that could use it. We had no TV ourselves at the time, so it was a bit of a temptation to just keep it for ourselves, but it had been entrusted to us to pass on, so we were happy to oblige. We knew of a poor village school and gladly gave it to them, along with a set of educational character-building video cassettes that the teachers could use with the children as part of their extra-curricular activities.

Donated TV to needy school

Toxic Refuge for Roma

We heard about a center for displaced Roma who had lost their homes. Their location was near the divided city of Mitrovica in the north of Kosovo. Not only were these poor people suffering from the effects of war, but the U.N. had chosen to put them near a defunct and abandoned car-battery factory that was polluted with toxic lead poisoning. Many of the Roma were getting sick from the contamination. When we showed up with a van full of aid for them, we were immediately circled by the residents. Distributing aid to the Roma is often difficult, as they generally do not have much etiquette and are not known for being very orderly. Oftentimes the Roma elder would have a list of the families and would call them by name to come and receive their aid. More often, it would end up turning into a free-for-all with people pushing and shoving and grabbing for whatever they could get. We had two bundles of lovely, hand-knitted "Peace Dolls" (200 in all) that we had received from a women's group in the U.K. We wanted to make sure every family got a doll. In order to do that, Andrew had to climb up on the roof of the van and place a doll into the outstretched hands of the mothers below.

No Wood? No Worries!

In the midst of a particularly cold Balkan winter, the mother of one of our trainees came to our door crying and asking us for help. Her alcoholic husband was spending what little money they had on drink and would become abusive if she confronted him about it. They had a humble house with two rooms where seven people slept. We knew the family very well. Two of their daughters were our trainees, and we had often helped them with aid. We went for a visit. While there, we noticed that the house inside was extremely cold. They were also suffering because they had run out of firewood and had no money to buy more. During the old Yugoslav system, most of the heating had been supplied by electricity, but the overload of so much usage during this particularly cold period had caused the transformer at the end of their street to blow. There was little hope that it would be repaired any time soon. We gave them some of our wood, but that would not last them long.

A day or two later, one of our international police friends handed us 200 euros (almost $275 U.S. at this writing) and said he felt very sorry for how the local people were suffering. He wanted to help someone. Could we give that money to whomever we thought needed it most. We told our friends we had money for firewood and drove them in our van to pick it up. Coincidentally, they were living just across the street from this caring benefactor.

Over the years since then, we have kept in close contact with this precious family. When we visit Kosovo, we stay with them, and they treat us as honored guests. Since that time, they have been able to add another bedroom and a bathroom (they only had an outhouse before), the children have grown up and gotten good jobs, and the father was able to turn his life around and rid himself of his alcohol addiction.

Motivational Training of Young People

During our early days in Kosovo, one of our first projects was to supply motivational and leadership training in English to Serbian minority youth. Our group numbered up to fifty at times, and we met every Saturday for about three hours. These were high school students, ages sixteen through eighteen. While the schools themselves did the best job they could teaching the basic scholastic subjects, we felt young people in general were lacking in the type of training that would help them in the future to be strong, diligent, honest, hard-working leaders, willing to take initiative while maintaining high moral standards. I guess this could be true of youth everywhere in the world. We also believed that in today's world, a good, working knowledge of English would be beneficial in helping them find better, higher paying jobs. It's amazing now, after thirteen years, to look back and visit some of our trainees from that time and see them all grown up in their late twenties, married, working in good jobs, and successful.

Here are just a few examples:

- One Roma boy started his own school for Roma children, attended numerous international conferences on human and Roma rights, attended university in Switzerland studying human rights, became Kosovo's Roma representative to the U.N.
- One boy became vice-mayor of his hometown.
- One boy has worked for various international organizations, and is an entrepreneur in various cottage industries.
- One boy now directs his own humanitarian NGO (Non-Government Organization).
- One girl has worked for World Vision, a bank, and several private businesses. She is currently teaching.
- One girl operates her own shop.
- One girl is the director of a kindergarten.
- One boy is a grade school teacher in his home village.
- One girl married a Belgian man and is living in Luxembourg.

More than just their success, we have been impressed by the ability of these young people to manifest many of the positive attitudes that we worked so hard to instill in them: compassion for others, a willing spirit, giving sacrificially to help others with less, etc.

CHAPTER FIFTEEN

THOUGHTS ON THE MAYAN CIVILIZATION

As our children grew older, one by one they headed off to start their own lives. Our youngest daughter, Angelina, was the one who most manifested the traveling gene that she had inherited from her father and me. Her first adventure was in Equatorial Guinea, a small country in West Africa sandwiched between Cameroon and Gabon. From there, she went to Ukraine and Russia, where she met her boyfriend who would later become her husband.

They were invited to the picturesque Central American country of Guatemala to head up their own volunteer initiative where they stayed for three years. Angelina was due to have her first child (our first grandson), and I wanted to be a help to her in any way I could, as I knew she had a very busy schedule. It was Christmastime, and their team was visiting different juvenile centers, hospitals, and orphanages with a Christmas program that included serving hot meals for those in attendance. I have personally never seen so many orphans or AIDS patients in my life. It was so sad. Many children had lost their parents to AIDS. We met six beautiful young girls, sisters, who had just lost their single mother to AIDS!

Angelina's team gave educational materials to the orphanages.

One of their colleagues would go alone out into the jungle in a car loaded down with books, food, and clothes for the needy. Such an amazing work! Similar to India, Guatemala had countless travelers coming in search of "enlightenment" in the jungle because of the mystique of the Mayan culture. We ourselves visited Tikal, the site of one of the most famous Mayan pyramid ruins.

Guatemala is one of the most beautiful countries in the world, but there is also much crime and suffering because of drug trafficking. It was an incredible contrast of beautiful and ugly, good and bad. The country itself was like a garden, rich and lush. Situated in the highlands, the temperature was spring-like all year long. There were mountain lakes, with active volcanoes stretching forth in the distance, and sugar cane plantations down in the valleys. There were many tropical fruits we had never seen or tasted before, but there was also much poverty and suffering among the common people.

Guatemalan green market

CHAPTER 15: THOUGHTS ON THE MAYAN CIVILIZATION 131

The crime rate there is among some of the highest in the world. Simple shops had an armed guard standing at their entrance to discourage thieves, and mothers could not push their baby prams without a sheet covering the front for fear the baby might be kidnapped. Guatemala is on the main drug route to the U.S., and there were parts of Guatemala City where we were warned it was too dangerous for foreigners to go. But this stark contrast was no more visible than when comparing the peaceful indigenous Mayan population to the foreign tourists.

Angelina and Andrew at Lake Atitlan

While there we had a wonderful opportunity to visit the town of Panajachel that is set on the banks of the peaceful Lake Atitlan in the middle of what was once the center of the rich Mayan culture. The main event of the day for visitors to this serene setting is to watch the sun set behind the three volcanoes: San Pedro, Toliman, and Atitlan, which rim the west side of the lake. Here the pleasures of life are simple, like swimming where the volcanic hot springs rise up into the lake, creating a curious mixture of ice-cold, tepid, and very hot water. During our two days there it was a curious sight—tourists on temporary reprieve from their fast-paced, pressure-driven, "civilized" lives in juxtaposition with Mayan women peacefully weaving their beautiful multicolored cloth, some with an infant child, grandchild, or perhaps even great-grandchild sleeping peacefully in a sling across their backs or playing quietly nearby. What a contrast!

Guatemalan woman weaving

Although the Mayans of today must sell their wares to the foreigners in order to survive, they have not let the "world" pollute their lives. Industrious and hard-working, they labor in rhythm, sunrise to sunset, producing lovely traditional garments. Not swept along by the fashion trends, they proudly wear their own creations, as do their children. They are not dependent on the pharmaceutical system but find the medicines they need right in the trees and plants that grow locally. They even produce their own cosmetics from the same natural organic substances that their ancestors had used centuries before. My hair did better with the herbal shampoo

I bought from them than it did with any commercially-produced shampoo, herbal or otherwise, that I have ever tried,

While sitting beneath the softly waving palm trees, listening to gentle waves lapping the lakeshore and observing the sun setting behind the volcanoes, I felt like I had been taken up to God's heavenly kingdom and shown the reason that I had been created —to enjoy it all!

Jet-setters, get-aheaders, and others consider the Mayans to be backward country folk, but I'm not so sure. Speed and stress take the joy out of life, but slowing down and bringing our priorities into line with what God intended when He designed it all puts it back. One thing I know, there was definitely something special about the culture that produced these peaceful, loving people.

CHAPTER SIXTEEN

PERU

A colleague we had known for years in Romania and Serbia had returned to Peru when his Peruvian wife needed special medical attention. When he heard we were in Guatemala, he wrote to us asking if we could come and help him. Peru is a poor country. The capital, Lima, is on the Pacific coast and is ringed with large, sprawling shantytown areas that climb ever upward into the dry, dusty hills that form the foothills of the Andes Mountains towering above. Like so many other poverty-stricken countries, tens of thousands of village peasants migrate annually to the big cities in search of employment and some way to support their families. Sadly, in Lima's case, most of these poor migrants end up in the shanty towns, able to barely scratch out a living.

A Peruvian friend had taken us up to a particular area in one of the shanty towns where he had been conducting children's activities once a week and also giving adult Bible training. We did some children's shows together, but there was nowhere on the rocky terraced hillside to congregate except in the narrow mud-packed road. Since one of our sponsors in the U.S. had wanted to come to Peru to help with a humanitarian construction project,

Shanty area children's program

Community center construction

we came up with the idea to build a small wooden community center for the residents of this area right there among the shanties. Our supporter agreed and came with a team of six to do the work. They stayed with us over two weeks, bought all the materials out of their own pockets, and left us with a generous gift when they left. From then on, we had a nice, cozy venue in which to continue our activities with those sweet, humble people. The Rotary Club had invited us to help them with a medical camp that they would be conducting in one of the poorer barrios of Lima. The residents lined up for quite a distance waiting for their turn to receive general check-ups, eye or dental examinations, or have a particular malady diagnosed. The Finnish government had donated a mobile eye clinic to the Rotary Club. To help the children not get fidgety, we had all sorts of fun activities for them. Both the parents and the organizers were very thankful for our presence.

Table at the medical camp

Andrew with mobile eye clinic van

About twenty-five kilometers outside Lima, alongside a lovely mountain stream that flowed down from the Andes, was a home for about 120 boys who were either rescued from drug-addicted parents, or were drug users themselves. A kind-hearted Canadian man had founded this center some twenty-five years before and had been running it almost single-handedly with the help of two middle-aged women and a few local employees. Work therapy was one of the things that kept the boys occupied and focused during the day as part of their rehabilitation process. In the evening, it was music. Lots of music, and drums, lots of drums! We staged a performance for them, and then they entertained us. It was great. Once Latinos start playing music, it can go on for hours! We had a real heart for these boys and truly respected the work that was being done by the staff there, as they really were changing the world.

Peru is comprised of three distinct regions: coastal desert, mountainous, and jungle. The mountainous and jungle regions are especially noted for the various indigenous groups and peoples,

Anne with indigenous Peruvian women

many of whom are discriminated against or do not receive proper care or recognition from the government. During an international summit that was being held in Lima, these groups were planning protests, but the government did not want them disturbing the official proceedings and so set up a "protest area" in another part of the city for them. We wanted to meet some of these people and have a chance to stand in solidarity with them, and so we went to the protest area, even though some of our friends had warned us that it might be dangerous.

Following is the story of a Peruvian Catholic priest who has also laid down his life to help the poor Lima street boys who come down into the city each day from the shanty-towns:

"If You Build It, They Will Come" –
The Story of Padre Luis Cordero

Padre Cordero was enjoying his final minutes of solitude as he finished his lunch at a high-class restaurant in the most affluent suburb of Lima, Peru. As he reflected on his afternoon commitments, he was interrupted from his ponderings by a raggedy boy who had been lurking in the entranceway of the restaurant and had finally mustered the nerve to approach his table. No, the brownish boy was not selling sweets or cigarettes. He asked a simple question: "Sir, if you are not going to eat the rest of that lettuce on your plate, do you mind if I do?" His heart-strings plucked by the boy's obvious need, as well as his child-like faith to ask, Padre Cordero immediately took the boy through the line, inviting him to take whatever he liked onto his tray.

Ignoring the looks of disapproval from the waiters in attendance, he led the boy to his own table. But instead of sitting down and diving into the feast before him, the boy instead ran quickly out of the restaurant. Dismayed by the urchin's erratic behavior, the kindly Cordero waited to see how things would develop. The boy reappeared a few minutes later, this time with two bedraggled boys trailing sheepishly behind him. He had wanted to share his good fortune with his friends. As they ate, the boys explained that they came from very poor families that lived in the dusty shantytowns on the outskirts of Lima that had sprung up over the years as families flocked from the mountain villages and Amazon regions to the big city in search of employment. The boys tried to help their families survive by washing the cars in the parking lot, in hopes of getting a tip when the owners came out. Face to face with such need, Luis Cordero felt God speaking to his heart. His life would never be the same after that day.

With Padre Cordero

The vision that Cordero received that day was to set up a center for boys—a place where they could escape the rigors of life that weighed too heavily on them at such a tender young age. The problem was, he had nothing to start with—no property, no buildings, no money. It was at this point that Luis Cordero became who we shall call the "Peruvian George Müller." (See note, next page.)

In answer to his prayers, the mayor of a Lima suburb offered Cordero a 5,000 m^2 plot of municipal land (approximately 1.25 acres) on which to build. As word spread, generous sponsors began to offer building materials, and, due to a series of miracles, a school

> **George Müller (1805-1898) Biography excerpt:** "Three weeks after their marriage, [George and wife Mary] decided to depend upon God alone to provide their needs. They carried it to the extent that they would not give definite answers to inquiries as to whether or not they were in need of money at any particular moment. At the time of need, there would always seem to be funds available from some source... No matter how pressing was the need, George simply renewed his prayers, and either money or food always came in time to save the situation."

and other buildings for the children were soon built. Not everything, however, went according to plan. Being in a rather affluent area, some of the neighbors whose land bordered Cordero's soon began to complain about the scruffy, ragtag children who were filtering down into their neighborhood from the poverty stricken areas. They filed a petition with the municipality, and the Padre was forced to build a high wall around the property, with a three-meter (ten-foot) "no man's land" between the wall and neighboring property. Cordero did not want his refuge for the boys to become a fortress, but he soon realized that the wall would be for their benefit, as it would protect the boys from the outside world and give them an increased feeling of freedom to be themselves as they studied and played. To complete the wall, he would need 23,000 bricks. A friendly factory owner offered him 1,000 bricks, a start to be sure, but where would he get the rest? At that moment, an irate customer stormed into the office expressing his discontent with an order of bricks he had just received and insisting that they all be returned. The number of bricks: 23,000!

The miracle that became the **Roncalli del Peru** began in 1987 and continues today with Padre Cordero (now 86 years old) still overseeing the day-to-day operations. Roncalli is living proof of God's power to supply in answer to simple faith and fervent prayer.

Boys from Roncalli Institute   *Receiving humanitarian aid*

After two and a half wonderful years in Guatemala and Peru, we realized that this respite from Eastern Europe was now coming to an end. We had been refreshed, rejuvenated, and re-envisioned, and now the republics of the former Yugoslavia, where we had called home for twelve years, were once again calling to us to return.

I had always enjoyed our ocean travels, but my heart's desire had always been to take a real cruise. Andrew told me if I wanted to take a cruise back to Europe, I would have to search online and find one that was convenient for our time schedule and fit into our budget. I began to investigate the cruise industry and found that there were two distinct seasons depending on the time of year, the Mediterranean season and the Caribbean season, and that the cruise ships would transition from one area to the other when the season was changing. These transatlantic transitional cruises were offered at huge discounts in order to fill the ships as much as possible to cover costs.

I could not find any leaving from Peru, but I did find one leaving from Brazil. Unfortunately for us, even though the cost of the cruise was very low, the cost of getting to Brazil from Peru was prohibitively high. I found another cruise leaving from Fort Lauderdale, Florida, for about the same price, and I found discount airfares to fly from Lima to Fort Lauderdale. Since our son was living in the Miami area, we decided to go this route. We would be able to visit him, his wife, and our two granddaughters, aged nine and seven. We would then board ship for the cruise. The first six days would be spent calling on ports in the Caribbean before crossing over to Italy where our daughter, Angelina, and her family were then living. We would have loved to also see Joy, but that would have to wait, as she was then living in Sweden and working as a financial analyst/blogger.

It is always a good time to reflect on the past and prepare for the future when moving from one field to another. Taking this cruise was the chance of a lifetime for us! In his farewell address to the passengers, the captain of our vessel said that never before in his career had he experienced such a calm transatlantic crossing. I guess he did not realize he had God's ambassadors aboard!

We picked back up with our humanitarian activities in Serbia and Kosovo, where we continue to this day.

Ready to sail on Costa Fortuna

CHAPTER SEVENTEEN

Conclusion – The Simple Life

I heard a story some time ago that was very poignant and summarized quite well the state of the modern world and how so few people are able to fully enjoy the best things in life because of their over-emphasis on money and possessions. It goes something like this …

A wealthy American was taking his vacation in Mexico, and while standing on the pier of a small coastal village, he observed a humble fishing boat docking. Inside the boat were several large yellow-fin tuna. The man complimented the Mexican on the quality of his fish and asked how long it had taken him to catch them.

"Only a little while," replied the fisherman.

The rich American then asked why he didn't stay out longer so he could catch more fish. The Mexican answered that what he had caught was enough to support his family's immediate needs.

The American then asked, "But what do you do with the rest of your time?"

The fisherman responded, "I sleep late, fish a little, play

with my children, take siestas with my wife, Maria, stroll into the village each evening where I sip wine and play guitar with my amigos. I have a full and busy life, señor."

The American scoffed, "I'm an investment banker with an MBA, and I could help you. You should spend more time fishing and, with the proceeds, buy a bigger boat. With the proceeds from the bigger boat, you could buy several boats. Eventually, you would have a fleet of fishing boats. Instead of selling your catch to a middleman, you would sell directly to the processor, eventually opening your own cannery. You would control the product, processing, and distribution. Of course, you would need to leave this small coastal fishing village and move to Mexico City, then Los Angeles, and eventually New York City, where you will run your expanding enterprise."

The Mexican fisherman calmly asked, "But, how long will this all take?"

The American replied, "If you work hard, within fifteen to twenty years."

"And then?" asked the Mexican.

The American laughed and said, "When the time is right, you would sell your company stock to the public and become very rich. You would make millions!"

"Millions, señor? Then what?"

The American said, "That's the best part! Then you could finally retire, move to a small coastal fishing village where you could sleep late, fish a little, play with your grandchildren, take siestas with your wife, stroll to the village in the evenings where you could sip wine and play your guitar with your amigos."

When living in the now, it's hard to see sometimes what you are actually accomplishing toward the overall picture. Like a thread in a tapestry, only when you look back at the finished product can you see how important each thread has become to the whole.

I heard the story of a missionary who in the late 1800s had returned by ship from Africa and was arriving in New York harbor. On that same ship was Teddy Roosevelt, who himself was returning from a diplomatic tour of Europe. People on the shore were waving, and bands were playing as the President walked down the gangplank. As the crowd dissipated, no one had come out to meet the missionary. His first thought was one of discouragement, and he asked the Lord, "Why hasn't anyone come out to welcome me home?" To which the Lord replied, "Because you are not home yet!"

I became a "fisher of men" many years ago when I decided to live for others and walk as Jesus walked on the earth. His friends were mostly the outcasts of society, the harlots and tax collectors —the "sinners." He had long hair, did not own his own home, blessed the homes of those he stayed with, helped the poor and needy, rejected the self-righteous, hypocritical religious system of his day, went up into mountains to escape the spirit of the world, and changed everything. He is the ultimate role model. (See 1 Corinthians 11:1.)

The advertising industry would have you believe that you must have the most beautiful skin, the most slender figure, the newest car, or the fastest computer. These are the things people should strive for (you can continue that list *ad infinitum*). Did you know that the longest living humans on the face of the earth, the societies with the highest percentage of centenarians, are the ones who live high in the mountains away from most modern conveniences? They walk, they do manual labor in the fields, they have close family ties, and they live their lives like God intended man to live. In simplicity! (See 2 Corinthians 1:12, 11:3.)

The journey through life is always mixed with many twists and turns, ups and downs, highs and lows, potholes and smooth sailing ahead. The path we choose to take will either lead us through the "strait and narrow gate" or down the "broad way" (Matthew 7:13-14). It all depends on our own decisions, decisions we are faced with hundreds of times each day. I am looking forward to that day when my job here on earth is done, and my Father welcomes me "home" with that approving statement: "Well done, thou good and faithful servant."

My prayer is that after reading the stories and testimonies in this book, and seeing how amazingly the Lord has worked in my life, that it will inspire you, too, to take that step of faith and trust Him to rule in your life as well. You can know God's Word, but it's when you start putting it into action that it will take effect, and you will begin to see His Power manifested in your life.

If you would like to contact Anne Ranta or learn more about her charitable activities, please visit:

www.HealingHeartsBalkans.org

About The Author

Anne Ranta

Anne Ranta, a post-World War II baby boomer, was born in Finland. Sadly, she lost her mother through complications during her delivery, due primarily to malfeasance on the part of the attending physician. With her father working hard as a foreman in a paper factory, Anne was an only child raised primarily by her grandmother with the help of several aunts.

She majored in social anthropology at university in Sweden, and it was there that her desire to travel and experience the world was born. Her quest began with a field trip to Israel and the Palestinian territories where she studied the Bedouin culture. But it was back in Sweden when some Christian young people came to her campus singing and witnessing that her life truly changed and she began to hear God's call on her life.

She met her husband, Andrew, while traveling in southern Europe, and together they began an adventure that has crossed five continents and continues to this day. Their three children were raised in foreign lands and were taught solely by home schooling while living out extraordinary experiences that most children will only ever be able to read about in a textbook. They have four grandchildren.

Over the years Anne has written numerous essays documenting her life and travels, many of which have been printed in different periodicals. ***Unplugged from the Norm*** is a non-fictional account of her life, travels, and service for the Lord over the past 40 years.